AF703920

The Political Economy of the Space Age

How Science and Technology Shape the Evolution of Human Society

Andrea Sommariva

Vernon Series in Economic Development

VERNON PRESS

Copyright © 2018 Vernon Press, an imprint of Vernon Art and Science Inc, on behalf of the author.

All rights reserved. No part of this publication may be reproduced, stored in a retrieval system, or transmitted in any form or by any means, electronic, mechanical, photocopying, recording, or otherwise, without the prior permission of Vernon Art and Science Inc.

www.vernonpress.com

In the Americas:
Vernon Press
1000 N West Street,
Suite 1200, Wilmington,
Delaware 19801
United States

In the rest of the world:
Vernon Press
C/Sancti Espiritu 17,
Malaga, 29006
Spain

Vernon Series in Economic Development

Library of Congress Control Number: 2018933283

ISBN: 978-1-62273-264-7

Cover design by Vernon Press, using elements by Nasa

Product and company names mentioned in this work are the trademarks of their respective owners. While every care has been taken in preparing this work, neither the authors nor Vernon Art and Science Inc. may be held responsible for any loss or damage caused or alleged to be caused directly or indirectly by the information contained in it.

Table of Contents

List of Figures *ix*

List of Tables *xi*

Acronyms *xiii*

Forward *xvii*

Introduction *xix*

Chapter 1 The invention stage 1

1.1 K. E. Tsiolkovsky 1
1.2 R. H. Goddard 4
1.3 H. J. Oberth 7
1.4 P. Korolev and W. von Braun 10
1.4.1 The Soviet Union: Sergei Pavlovich Korolev 10
1.4.2 Germany: Wernher von Braun. 13
1.5 Obstacles to change 16
1.5.1 The United States 17
1.5.2 The Soviet Union 19
1.5.3 Germany 20
1.6 Lessons from history 22
Appendix - The Rocket Equation 24
Bibliography 27

Chapter 2 The dawn of the space economy, 1950-1970: the innovation process stage 29

2.1 Military programs 30
2.2 The space programs 33
2.2.1 The Moon project 35
2.2.2 The government satellites' programs 40
2.3 The dawn of commercial satellite activities 43
2.4 The political economy of the space age during the 1950s and the 1960s 45
Appendix - Von Braun Mars mission: the missed opportunity 48
Bibliography 52

Chapter 3 The space economy between 1970 and today: the diffusion stage 53

3.1 The commercial space sector 53
3.1.1 The commercial space service sector 54
3.1.2 The commercial space infrastructure and support industries 56
3.1.3 Recent policies changes 60
3.1.4 Launching industry's developments 61
3.2 Public administrations engaged in space activities 63
3.2.1 Space agencies 64
3.2.2 Activities of space agencies 67
3.3 International collaboration 74
3.3.1 International space institutions 74
3.3.2 Joint institutional space programs 76
3.4 Impacts on the world economy 78
3.4.1 Cost/efficiency gains 78
3.4.2 Economic impact of space exploration 82
3.5 New private space initiatives 83
3.6 The political economy of the old space age 86
Bibliography 88

Chapter 4 The political economy of the new space age 89

4.1 Why governments should dedicate significant public resources to going into space 90
4.1.1 Economic benefits 90
4.1.2 Political benefits 101
4.2 Whether present institutions are adequate to the uses of space resources 103
4.2.1 The United States 104
4.2.2 Europe 106
4.2.3 China 108
4.2.4 Japan 109
4.2.5 Russia 111
4.2.6 India 112
4.2.7 The prospects of international collaboration in space 113
4.3 International legal framework 115
4.4 Conclusion 117
Bibliography 120

Chapter 5 The space economy beyond Earth orbit: space mining 121

5.1 Low costs of accessing space 121
5.2 Space agencies programs 123
5.3 Global problems 126
5.4 The Moon's mineral resources 127
5.5 The asteroids' mineral resources. 130
5.6 Concepts of mining and processing 136
5.7 The economics of space mining 138
5.7.1 In-Space as a reference market 139
5.7.2 Earth as a reference market 141
5.8 Conclusions 145
Appendix - Technical Note 146
Bibliography 152

Chapter 6 The development of the space economy beyond Earth orbit: Mars 153

6.1 The construction of a permanent base on Mars and its sustainability. 154
6.1.1 Mars physical characteristics 154
6.1.2 Technical feasibility 157
6.1.3 The sustainability of the Mars base. 162
6.2 Human exploration of Mars 169
6.2.1 Propulsion 169
6.2.2 Trajectories 173
6.2.3 Shielding of cosmic radiation 177
6.2.4 *Descent/ascent vehicles* 179
6.2.5 Landing site 180
6.2.6 The Mars outpost base camp and logistics 181
6.3 Conclusions 185
Bibliography 187

Chapter 7 The ultimate challenge: the exploration and colonization of extrasolar planets. 189

7.1 How to go there: Einstein and the special relativity theory 189
7.2 The search for extrasolar planets capable of sustaining life 194
7.2.1 Methods to detect an extrasolar planet capable of sustaining life 194
7.2.2 Where antimatter propulsion will bring humanity 197
7.3 The exploratory and colonizing missions 206
7.3.1 Motivations for human missions 208
7.4 Overcoming the limits imposed by special relativity: myth or reality? 214

Bibliography 218

Chapter 8 Final Remarks 221

Bibliography 230

Index *231*

List of Figures

Figure 1.1 – K. E. Tsiolkovsky 2

Figure 1.2 – R. H. Goddard 5

Figure 1.3 – H. J. Oberth, von Braun, and other members of the VfR 8

Figure 1.4 – Korolev's Booking photo 1938 12

Figure 1.5 – V 2 14

Figure 2.1 – Design of von Braun space station in Colliers Weekly 33

Figure 2.2 – Apollo Space Flight Vehicles 37

Figure 5.1 – Map of the asteroids 131

Figure 5.2 – Mission to an Aten asteroid 148

Figure 5.3 – Mission to an Apollo type of Asteroid 150

Figure 6.1 – Rotating wheel 159

Figure 6.2 – Trajectories for conjunction and opposition Mars Missions 174

Figure 7.1 – Closest Star Systems to the solar system 197

List of Tables

Table 3.1 – Costs of the Heavy Launchers to geostationary orbit 61

Table 3.2 – Specific Impulses in Seconds 70

Table 3.3 – Selected economic effects of satellite applications 79

Table 3.4 – Private Space Initiatives 84

Table 3.5 – List of Space Investing Billionaires 85

Table 4.1 – Levelized costs of different power sources 92

Table 4.2 – Metric for some critical minerals in 2016 96

Table 4.3 – Demand for rare earth oxides in 2011 in metric tons 98

Table 4.4 – Summary of environmental sustainability metrics by PGM mine/project 100

Table 4.5 – Military and Space Agencies Expenditures (in billions of dollars) 102

Table 5.1 – Abundance of Chemicals and Metals on the Moon 128

Table 5.2 – Abundance of PGMs and Precious Metals in Asteroids 133

Table 5.3 – Products and processes 136

Table 5.4 – Δv Required for Lunar Transfer 140

Table 5.5 – Δv Required for Transfer of Minerals to Earth Orbit 142

Table 6.1 – Main characteristics of Mars and Earth 155

Table 7.1 – Possible targets for interstellar missions 199

Acronyms

AMAZE	Additive Manufacturing Aiming Towards Zero Waste & Efficient Production of High-Tech Metal Products
APRSAF	Asia-Pacific Regional Space Agency Forum
APSCO	Asia-Pacific Space Cooperation Organization
ARM	Asteroid Redirect Mission
AU	Astronautical unit
CASC	China Aerospace Corporation
CERN	European Organization for Nuclear Research
CNSA	China National Space administration Comsat Communications Satellite Corporation
COSPAR	United Nation Committee on Space Research
DBS	Direct Broadcast Satellite
DoS	Indian Department of Space
DRM	Design Reference Mission
DSP	Defense Support Program
EADS	European Aeronautic Defense and Space Company
ELT	Extremely Large Telescope
EOS	Earth Observing System
ESA	European Space Agency
EU	European Union
FAA	Federal Aviation Administration
GAO	Government Accountability Office
GEO	Geosynchronous Orbit
GDL	Leningrad Gas Dynamics Laboratory
GDP	Gross Domestic Product
GIRD	Moscow Group for Investigation of Reactive Motion
GNP	Gross National Product
GNSS	Global Navigation Satellite System
GOES	Geostationary Operational Environmental Satellite
GPS	Global Positioning System

HIAD	Hyper-sonic Inflatable Atmospheric Decelerators
HOPE	Human Outer Planets Exploration
HSF	Human Space Flight
IC	Integrated Circuits
ICBM	Intercontinental Ballistic Missile
ICF	Inertial Confinement Fusion
ICSU	International Council of Scientific Unions
IGY	International Geophysical Year
IISL	International Institute of Space Law
IRBM	Intermediate Range Ballistic Missile
ISAS	Japanese Institute of Science Astronautics
ISRO	Indian Space Research Organization
ISS	International Space Station
ISU	International Space University
ITS	Intelligent Technology Systems
ITU	International Telecommunication Union
JAXA	Japanese Space Agency
JIOA	Joint Intelligence Objectives Agency
JPL	Jet Propulsion Laboratory
JWST	James Webb Space Telescope
LEM	Lunar Excursion Module
LEO	Low Earth Orbit
LIP	Large Igneous Provinces
LOR	Lunar Orbit Rendezvous
MAD	Mutual Assured Destruction
MiDAS	Missile Defense Alarm System
MOM	Mars Orbiter Mission
MPD	Magneto-plasma Dynamic
MSL	Mars Science Laboratory
MTF	Magnetized target fusion
NACA	National Advisory Committee for Aeronautics
NAL	Japanese Aerospace Laboratory

NASA	National Aeronautics and Space Administration
NASDA	Japanese National Space Development Agency
NDRC	National Defense Research Committee
NEA	Near Earth Asteroids
NERVA	Nuclear Engine for Rocket Vehicle Application
NIAC	NASA Institute for Advanced Concepts
NISAR	NASA-ISRO SAR Mission
NPV	Net Present Value
OECD	Organization for Economic Cooperation and Development
OSRD	Office of Scientific Research and Development
OTP	The White House Office of Telecommunications Policy
PGM	Platinum Group Metals
PICA	Phenolic Impregnated Carbon Ablator
PLA	People Liberation Army
ppm	Parts per million
PS300	High Temperature Solid Lubricant Coating
RCA	Radio Corporation off America
REE	Rare Earth Elements
REM	Roentgen Equivalent Man
REO	Rare Earth Oxide
RFSA	Russian Federal Space Agency
RNII	Reaction Propulsion Scientific Research Institute
RPM	Resource Prospector Mission
SBPS	Space-based Solar Power
SLS	Space Launch System
SKA	Square Kilometer Array
SOI	Sphere of Influence
TESS	Transiting Exoplanet Survey Satellite
TRL	Technology Readiness Level
ULA	United Launch Alliance
UNCOPUOS	United Nations Committee on the Peaceful Uses of Outer Space

USAF	United States Air Force
USSR	Union of Soviet Socialist Republics
VfR	Verein für Raumschiffahrt
VISTA	Vehicle for Interplanetary Transport Applications

Forward

This book is dedicated to my friend Giovanni F. Bignami who has recently passed away. Nanni was an astrophysicist, fascinated by many other disciplines. He was a member of the Accademia dei Lincei and of the French Academy. He had key roles in the European Space Agency, and directed the largest institute for space research in France. He was elected Presidency of the COSPAR (Committee on Space Research World), was president of the Italian Space Agency and of the Italian National Institute of Astrophysics. This book was made possible by our long collaboration and discussions, not only on the technical aspects of space flight and astronautics, but also on the role of space exploration in the developments of our civilization. This book also benefited from discussions with several people at the International Academy of Astronautics, in particular, Prof. Giancarlo Genta and Art Dula. It is evident that all remaining errors are mine.

Introduction

Space and human space exploration have always fascinated mankind. The question of what is space has intrigued philosophers such as Lucretius, Kepler, and Kant; and scientists such as Newton, Mach, and Einstein. The concept of space is considered to be of fundamental importance to an understanding of the physical universe. Physical space was often conceived by scientists in three linear dimensions, although modern physicists usually consider it, with time, to be part of a boundless four-dimensional continuum known as spacetime. But disagreement continues between philosophers over whether it is itself an entity, a relationship between entities, or part of a conceptual framework.

Space exploration inspired modern non fiction and fiction works. Non fiction works include Tsiolkovsky's famous book *"The Exploration of Cosmic Space by Means of Reaction Engines"*[1]. Fiction works span from Verne to Asimov, A. C. Clark, Fred Hoyle, and Italo Calvino. They are not only pleasant entertainment but also ways to expand the reader's imagination, allowing him to explore responses to future scientific developments and to speculate on how they might develop. In the second half of the 20th century, space exploration moved away from the realm of pure imagination. The launch of the Sputnik 1 in October 1957 and the first landing on the moon by the American Apollo program in July 1969 opened the way to the exploration of the solar system.

During this period, a space economy emerged. According to the OECD[2] *"the space economy is the full range of activities and use of resources that create value and benefits to human beings in the course of exploring, researching, managing, and utilizing space".* It is a mixed economy. Governments participate through space agencies and military programs, while the private sector activities are centered on low earth orbit (Earth's observation) and geosynchronous orbit (telecommunications). The space economy has been analysed using different approaches from the effects of public expenditures on economic growth to its medium term effects on productivity and growth. This book analyzes the space economy from a broader perspective. This prospective is heavily influenced by the pioneering work of Nelson and Winter[3] on the evolutionary theory of economic change. This approach is both holistic and evolutionary. It is holistic in that the parts of any whole cannot exist and be

[1]See Tsiolkovsky K. English edition (1995).

[2]OECD (2014)

[3]See R. R. Nelson et al (1982).

understood except in their relation to each other and the whole. It is evolutionary in the sense that it studies processes that transform the economy through the interactions among diverse economic agents, governments, and the extra systemic environment in which governments operate.

At the core of the discussion is the role of technical change in economic development. Such dynamic adjustments involve advancement in science and modifications in technology, changes in organization, and the introduction of new rules of behavior. As a result, new waves of consumption, technology, and socio-economic organization emerge overwhelming pre-existing practices. In the economics and management textbooks there exist a wide range of reference the Schumpeter's Trilogy of 'Invention-Innovation-Diffusion'. Schumpeter divides the technological change process into three stages. The first stage is the invention process, encompassing the generation of new ideas. The second stage is the innovation process encompassing the development of new ideas into marketable products and processes. The third stage is the diffusion stage, in which the new products and processes spread across the potential market.

The social process which produces inventions and the social process which produces innovations do not stand in any invariant relation to each other and such relation as they display is much more complex than appears. In many cases, invention and innovation are the result of conscious efforts to cope with a problem independently presented by an economic situation or certain features of it. Sometimes innovation is so conditioned, whereas the corresponding invention occurred independently of any practical need. Historical analysis indicates that new technologies often require a prolonged period of incubation before offering a wide range of opportunities for profitable new investments, new markets, and changes in socio-economic organization, in essence the evolution of human culture. But the same historical analysis indicates that these transitions periods are shortening. For example, it took several thousand years for agriculture to spread around the world, but only two centuries for industrialization to do so, and only a decade for the internet to spread globally. This acceleration indicates that technical change is becoming something contemporary and of immediate interest.

One important debate on technical change has centered on the relative roles of governments and/or other non profit institutions, and markets. An important conclusion of this broader approach is that the idea of government as a static bureaucratic organization only needed to fix market failures, leaving dynamic entrepreneurship and innovation to the private sector, is wrong, and that governments have important roles in fostering these changes. Governments, and in some cases other non profit institutions, have motives other than profit to invest in new technologies, determined by the internal and external socio-economic environment (the extra systemic environment) in which they operate. Once the opportunities for profitable new technologies are established, allocation of resources becomes the main mechanism for investments, and the market is the most efficient institution to allocate resources. But there is a growing perception in social sciences that, in a complex and non stationary environment, one cannot deduce behavior solely from knowledge of market-delivered information and self-seeking goals of individual agents. Public institutions shaping the vision of the world, behavioral conventions, and interactions between economic agents are important in explaining what economic agents do, what kind of technical progress they expect in the future, how much they invest in innovations, what appropriability mechanisms they build, and how much they cooperate and compete with each other. In summary, laws, policies, and institutions are an important part of the environment that shapes the evolution of private sector activities. As the above factors and the extra systemic environment undergo continuing changes, over the long run the cumulative result of private and public actions and reactions gradually modify the basic structure of society.

The development of the space economy is a good case study to verify if this broader approach fits the historical evidence. It may also provide insights for public policies directed to foster future evolution. This book examines: (i) the history of space exploration and of the space economy from its beginning up to now, and the respective roles of governments and the private sector in the evolution of the space economy; (ii) short-medium-term prospects of human deep space exploration and the expansion of the space economy beyond Earth's orbit, and their interrelations; and (iii) possible longer-term developments.

History helps to better understand motivations and constraints - technical, political, and economical - that shaped the growth of the space economy. Short-medium term prospects enable to identify the driving forces that will shape the next phase in the evolution of the space economy beyond Earth's orbit. Central to this discussion are the global issues such as population changes, critical or limited natural resources, and environmental damages, which are transnational in nature, and threaten to affect all. The book explores how space technology could mitigate some of these problems and/or solve some of them, and the role of public institutions and the private sector in fostering those changes. The book analyzes whether present institutions and policies are adequate to foster short-medium term developments and to suggest possible alternatives.

Long term expansion of the space economy in the solar system beyond the Moon requires technologies that are not yet mature, while longer term interstellar exploration and colonization, although within present scientific and technological capabilities, call for advances in technologies that we do not yet have. The book examines the political and institutional framework required in the longer term. Longer term expansion of the space economy call for a civilization in which human beings see themselves as inhabitants of a single planet and global governance is conducted on a cooperative international basis. The establishment of a new international framework is a gradual process driven not only by changes in objective conditions but also by a shift of values or understanding. Changing global issues are part of the objective conditions, and space technology is part of their solutions. Enhanced role education will play a crucial role when broader questions of values, trade offs, and strategies are explored.

A little practical advice before leaving the reader to progress through the book. To ease reading the text, all technical details have been placed in the footnotes or in end-of-chapter appendices. Bibliographic references are collected at the end of each chapter. They are unnecessary to understand the text. However, they are necessary to the curious reader wanting to know more.

Bibliography

Nelson R. R. et al.: "*An Evolutionary Theory of Economic Change*", The Belknap Press of Harvard University Press, 1982.

Tsiolkovsky K.:"*Exploration of the Universe with Reaction Machines: Exploring the Unknown*", The Nasa History Series, NASA SP 4407, Washington D.C., 1995.

Chapter 1

The invention stage

"Earth is the cradle of humanity, but one cannot live in a cradle forever."

Konstantin E. Tsiolkovsky

Astronautics is the theory and practice of navigation beyond Earth's atmosphere. Newton established astronautics' mathematical basis in his treatise *"The Mathematical Principles of Natural Philosophy"*. They are embedded in his laws of motion and gravitation. Reactions in the spaceship's engine produce enormous pressures, which cause the expulsion of gas and/or radiation at high speed in the direction opposite to travel. It is this reaction force that pushes forward the engine and the spaceship attached to the engine.

Although Newton laid the mathematical foundations of astronautic long ago, astronautics became a science in its own right in the early twentieth century. During the first half of the twentieth century, there were also some advances in space technologies and engineering. This chapter tries to answer the following questions: (i) who developed the science and the technologies of space flight, and under what circumstances; (ii) what kind of institutions were established, who financed them, and how did they produce, structure, and organize knowledge; (iii) what were the main obstacles of the earlier experiments in space flight; and (iv) what geopolitical changes occurred at the end of the 1940s that created the conditions for the birth of the space economy.

1.1 K. E. Tsiolkovsky

Konstantin Eduardovich Tsiolkovsky was born Sept. 17, 1857, in Izhevskoye, Russia. At a young age, he caught scarlet fever and became hard of hearing. He was not admitted to elementary schools because of his hearing problem. As a reclusive home-schooled child, he passed much of his time by reading books and became interested in mathematics and physics. As a teenager, he began to contemplate the possibility of space travel, inspired by the fiction of Jules Verne. After falling behind in his studies, Tsiolkovsky spent three years attending college in Moscow where Russian cosmism proponent Nikolai Fyodorov worked, one of the main proponents of Russian cosmism. It was in this intellectual environment that Tsiolkovsky further developed his interests in space travel. At the end of his studies, he was accomplished in both science and mathematics. He became a teacher at Kaluga, Russia, where he spent the rest of his life.

Starting from 1883, Tsiolkovsky theorized many aspects of space flight. He published his most famous work, *"The Exploration of Cosmic Space by Means of Reaction Engines"*, in 1903. In this book, Tsiolkovsky derived the basic formula for rocket propulsion (see Appendix to this chapter). The equation had been derived earlier by the British mathematician William Moore in 1813[1]. But Tsiolkovsky is honored as being the first to apply it to the question of whether rockets could achieve speeds necessary for space travel. This formula calculates the final velocity of a rocket from the escape speed of the gases, and the initial (including propellant) and final (without propellant) masses of the spaceship. This equation also plays an important role in evaluating the economic feasibility of a space project, as will become clear in subsequent chapters. In other theoretical works, he studied gyroscopes and liquid fuel rockets; calculated the escape velocity from a gravitational field; and analysed the problem of the control of a rocket that moves between gravitational fields. Among his works are the designs for rockets with steering thrusters, multistage boosters, space stations, airlocks for exiting a spaceship into the vacuum of space, and closed-cycle biological systems to provide food and oxygen for space colonies. He never complemented his theoretical writings with practical experiments in rocketry. But his work greatly influenced space and rocket researches and experiments carried out in the Soviet Union and Europe.

Figure 1.1 – K. E. Tsiolkovsky
Source: http://petersrussia.blogspot.it

Tsiolkovsky elaborated the theory

[1]See W. Moore (1813)

of space flight as a supplement to his philosophical inquiries on the cosmos. He is remembered for believing in the dominance of humanity throughout space, also known as anthropocosmism. Russian cosmism appeared as sort of antithesis to the classical physical paradigm based on strict differentiation of man and nature. Cosmism made an attempt to revive the ontology of an integral vision to unites man and the cosmos. These problems were discussed both in the scientific and the religious[2] form of cosmism. But while religious cosmism was more notable for the fantastic and speculative character of its discourses, the natural scientific trend, solving the problem of interconnection between man and cosmos, paid special attention to the comprehension of scientific achievements that confirmed that interconnection. Kholodny attempted to reconsider the question of man's place in nature and of his interrelations with the cosmos on the basis of natural scientific knowledge. In his opinion, the anthropocentric worldview started to collapse under the influence of the achievements of science and philosophy.

Tsiolkovsky was influenced by the works of N. G. Kholodny. In Kholodny's book *The Unknown Intelligence,* he propounded a philosophy of panpsychism. Panpsychism is the view that consciousness, mind or soul (psyche) is a universal and primordial feature of all things[3]. He believed humans would eventually colonize the Milky Way galaxy, because of the advance in science and industry. Although Tsiolkovsky's philosophical theories are now generally discarded, he was the first to inquire into motivations of man traveling into deep space, and to advance ideas on space stations, industrialization of space, and exploitation of space resources, which constitute actual programs of today space agencies and private space industries.

---

[2]N. Fedorov was the most significant author. He was a Russian Orthodox Christian philosopher, who was part of the Russian cosmism movement. Fedorov argued that evolutionary process was directed towards increased intelligence and its role in the development of life. Humanity is the culmination of evolution, as well as its creator and director. Humans must therefore direct evolution where their reason and morality dictate. Fedorov also argued that mortality is the most obvious indicator of the still imperfect, contradictory nature of humanity and the underlying reason for most evil and nihilism of humankind. Fedorov stated that the struggle against death can become the most natural cause uniting all people of Earth, regardless of their nationality, race, citizenship or wealth (he called this the Common Cause).

[3]Panpsychism is one of the oldest philosophical theories, and has been ascribed to philosophers like Thales, Parmenides, Plato, Averroes, Spinoza, Leibniz and William James. Panpsychism can also be seen in ancient philosophies such as Stoicism, Taoism, Vedanta and Mahayana Buddhism. During the 19th century, panpsychism was the default theory in philosophy of mind, but it saw a decline during the middle years of the 20th century with the rise of logical positivism

1.2 R. H. Goddard

Robert H. Goddard was an American engineer, professor, physicist, and inventor who is credited with designing and building the world's first liquid-fueled rocket. Goddard was born in Worcester, Massachusetts in 1859. With the introduction of electric power in American cities in the 1880s, the young Goddard became interested in science, specifically engineering and technology. At the age of sixteen, reading H. G. Wells' science fiction classic *The War of the Worlds,* he became curious about space traveling. Goddard enrolled at Worcester Polytechnic Institute in 1904 and received his B.S. degree in physics in 1908. In 1909, he began his graduate studies at Clark University, where he received his M.A. degree in physics in 1910, and then stayed at Clark to complete his Ph.D. in 1911. In 1912, he accepted a research fellowship at Princeton University's Palmer Physical Laboratory.

By 1913, he formulated the mathematics that allowed him to calculate the position and velocity of a rocket in vertical flight, given the weight of the rocket and of the propellant, and the velocity of the exhaust gases. As there is no evidence that Goddard knew the details of Tsiolkovsky's works, both he and Tsiolkovsky can share the title of Father of Astronautics. After this groundbreaking work, his first goal was to build a rocket to study the atmosphere. Not only would such investigation aid meteorology, but it was necessary to determine temperature, density and wind speed in order to design efficient space launch vehicles. He was very reluctant to admit that his ultimate goal was to develop a vehicle for flights into space, as most scientists in the United States did not consider such goals to be realistic or practical scientific pursuit, nor was the public yet ready to seriously consider such ideas.

At the Clark University, Goddard conducted static tests of solid-fuel rockets to measure their thrust and efficiency. He verified his earlier estimates that these rockets were converting only about two percent of their fuel into thrust. At this point he applied de Laval nozzles[4], which were generally used with steam turbine engines, which greatly improved efficiency. By mid-summer of 1915, Goddard obtained an average efficiency of forty percent with a nozzle exit velocity of 2,051 meters per second. Later on, connecting a combustion chamber full of gunpowder to various converging-diverging expansion nozzles, he was able to achieve in static tests engine efficiencies of more than sixty three per cent and exhaust velocities of over 2,134 meters per second.

---

[4]A de Laval nozzle nozzle is a tube that is pinched in the middle, making a carefully balanced, asymmetric hourglass shape. It is used to accelerate a hot, pressurized gas passing through it to a higher supersonic speed in the axial (thrust) direction, by converting the heat energy of the flow into kinetic energy. Because of this, the nozzle is widely used in some types of steam turbines and rocket engine nozzles.

In late 1919, the Smithsonian published Goddard's groundbreaking work, *A Method of Reaching Extreme Altitudes*[5]. The book describes Goddard's mathematical theories of rocket flight, his experiments with solid-fuel rockets, and the possibilities of exploring the Earth's atmosphere and beyond. Though most of this work dealt with the theoretical and experimental relations between propellant, rocket mass, thrust, and velocity, a final section - entitled "Calculation of minimum mass required to raise one pound to an "infinite' altitude" - discussed the possible uses of rockets, not only to reach the upper atmosphere, but to escape Earth's gravitational field. He determined that a rocket with an effective exhaust velocity of 2,134 meters per second and an initial weight of 281 kilograms would be able to send about half a kilogram of payload to Earth orbit.

By 1926, Goddard had constructed and successfully tested the first rocket using liquid fuel. Indeed, the flight of Goddard's rocket on March 16, 1926, at Auburn, Massachusetts, was as significant to the history of space flight as that of the Wright brothers at Kitty Hawk. He determined that fins alone were not sufficient to stabilize the rocket in flight and keep it on the desired trajectory in face of winds aloft and other disturbing forces. He added movable vanes in the exhaust, controlled by a gyroscope, to steer his rocket. He also introduced the more efficient swiveling engine in the rocket, basically the method used to steer large liquid-propellant missiles and launchers today. Goddard eventually relocated to Roswell, New Mexico, in summer of 1930, where he worked with his team of technicians in near-isolation and relative secrecy for years.

Figure 1.2 – R. H. Goddard
Source: https://www.rocketry.worldpress.com

[5]See Goddard (1946)

In 1934, he began work on his A series of rockets, 4 to 4.5 meters long, and powered by gasoline and liquid oxygen pressurized with nitrogen. The gyroscopic control system was housed in the middle of the rocket, between the propellant tanks. In 1936–1939, Goddard began work on the K and L series rockets, which were much more massive and designed to reach very high altitude. Goddard experimented with many of the features of today's large rockets, such as multiple combustion chambers and nozzles. In November, 1936, he flew the world's first rocket (L-7) with multiple chambers, hoping to increase thrust without increasing the size of a single chamber. This flight demonstrated that a rocket with multiple combustion chambers could fly stably and be easily guided.

As an instrument for reaching extreme altitudes, Goddard's rockets were not very successful; they did not achieve an altitude greater than 2.7 kilometers, while a balloon sonde had already reached 35 kilometers in 1921. Simply reaching high altitudes was not his primary goal. He tried, with a methodical approach, to perfect his liquid fuel engine and subsystems such as guidance and control, so that his rockets could eventually achieve high altitudes without tumbling in the rare atmosphere, providing a stable vehicle for the experiments it would eventually carry. He built the necessary turbo pumps and was on the verge of building larger, more reliable rockets to reach extreme altitudes when the Second World War intervened and changed the path of American history.

Goddard's pace was slower than the Germans' because he did not have the resources they did. Initially, Goddard's rocket research was funded out of his modest teaching salary. By 1916, he began to solicit potential sponsors for financial assistance, beginning with the Smithsonian Institution, the National Geographic Society, and the Aero Club of America. In January 1917, the Smithsonian agreed to provide Goddard with a five-year grant totaling 5,000 dollars. Afterwards, Clark University was able to contribute 3,500 dollars and the use of their physics laboratory for the project. In the spring of 1930, financier Daniel Guggenheim agreed to fund Goddard's research over the next four years for a total of 100,000 dollars. The Guggenheim family, especially Harry Guggenheim, would continue to support Goddard's work in the years to come.

In general, there was a lack of vision and serious interest in the United States concerning the potential of rocketry, especially in Washington. Although the Weather Bureau was interested in Goddard's rocket for atmospheric research, the Bureau could not secure governmental funding. Goddard believed his rocket research could be applied to many different military applications. He made proposals to the Navy and Army. No record exists in his papers of any interest by the Navy to his inquiry. However, Army Ordnance showed interested. He met several times with Army personnel, but no actions followed. The private sector also failed to see the potential of rocketry and its applications.

Goddard's liquid-fueled rocket was neglected by his country, according to aerospace historian Eugene Emme[6], but was noticed and advanced by other nations, especially Germany. Most of the United States largest universities were also slow to realize rocketry's potential. In academia, most people thought that liquid fuel was too expensive and dangerous. He was frequently ridiculed in the press since his publication by the Smithsonian in 1919. Although Goddard's discussion of targeting the Moon was only a small part of this work and was rather an illustration of the possibilities than a declaration of intent, the press sensationalized his ideas to the point of misrepresentation and ridicule.

Although his work in the field was revolutionary, it was only after his death in 1945 that he was recognized as one of the founding fathers of modern rocketry and astronautics. On Sept. 16, 1959, the 86th Congress authorized the issuance of a gold medal in his honor. He not only recognized the potential of rockets for atmospheric research, ballistic missiles and space travel, but he scientifically studies, designed, and constructed rockets needed to implement those ideas.

1.3 H. J. Oberth

Hermann Julius Oberth was born in 1894 in Hermannstadt, Austria-Hungary. At the age of eleven,when reading the writings of Jules Verne, especially From *the Earth to the Moon* and *Around the Moon,* he became fascinated with astronautics. Influenced by Verne's books and ideas, Oberth constructed his first rocket model the age of fourteen. He studied medicine in Munich, but his education was interrupted by service in the Austro-Hungarian army during World War I. After being wounded in the war, he found time to pursue his studies in astronautics. He performed experiments to simulate weightlessness and worked out a design for a long-range, liquid-propellant rocket that his commanding officer sent to the War Ministry. The design was rejected as a fantasy.

After the war, Oberth sought a Ph.D. degree at the University of Heidelberg with a dissertation based on his rocket design. It was rejected by the university in 1922, but Oberth partially underwrote publishing expenses, and it was published in 1923 with the title *Die Rakete zu den Planetenräumen*[7] (The Rocket into Interplanetary Space). The book discussed the feasibility of hu-

---

[6]See Eugene Emme (1965)
[7]See H. Oberth (1923, English edition 2014).

man spaceflight, laid out the basic equations of rocketry, and described how liquid propellants could vastly exceed the performance of solid-fuel rockets. The book gained Oberth widespread recognition. Until 1922 he was unfamiliar with the work of Robert Goddard, and, until 1925, with that of Konstantin Tsiolkovsky. After corresponding with both of them, he acknowledged their antecedence in deriving the equations associated with space flight. Oberth's *Wege zur Raumschiffahrt* (*Ways to Spaceflight*[8]) published in 1929 anticipated by thirty years the development of electric propulsion and of the ion rocket. In 1931 Oberth received a patent for a liquid-propellant rocket from the Romanian Patent Office, and the first liquid fuel rocket was launched on May 7, 1931, near Berlin.

Figure 1.3 – H. J. Oberth, von Braun, and other members of the VfR
Source: htts://eulegromania.ru

In 1923, Oberth became a member of the *Verein für Raumschiffahrt* (VfR) - the "Spaceflight Society" - an amateur rocketry group that had taken great inspiration from his book. He acted as a mentor to the enthusiasts who joined the Society, inspiring a young von Braun to apply himself to mathematics and

[8]An English translation of Oberth's 1929 book, *Ways to Spaceflight,* was published by NASA in 1972.

physics to better understand the rocket equations. A few years later, von Braun joined Oberth in the German Society of Space Travel, of which the older engineer had become president. This group successfully launched two liquid fuel rockets[9] that reached heights of over 2.4 kilometers.

Space travel and rocketry gained popularity in Germany following the publication of Herman Oberth's books. During the twenties, the VfR received funding from various private sources. Max Valier, one of the founders of the VfR, assisted in Fritz von Opel's solid-fuel rocket publicity stunts for the Opel company. In 1929, Hermann Oberth hoped to receive funding from Lang for a real life experimental rocket launch coinciding with Lang movie premiere *Frau im Mond* (The Woman in the Moon). But the incipient world depression dried up the private funding.

In September 1930, the VfR was contacted by the German army offering funding to the VfR. Rockets were one of the few fields of military spending not restricted by the Versailles treaty. In the Spring of 1932, the German Army issued a contract for a demonstration launch. Wernher Von Braun was in favor of the contract. The VfR eventually rejected the proposal and the dissension caused among its members contributed to the society dissolving itself in January 1934. With the advent of Hitler to power in 1933, rocketry almost immediately moved onto the national agenda. A larger facility was erected at the village of Peenemünde in northern Germany on the Baltic Sea, with von Braun as technical director.

During the Second World War, Oberth joined his former student von Braun, in developing and constructing the V-2 rocket. The 12,250 kilograms rocket carried warheads at over 3,500 mph. In 1950, Oberth moved on to Italy, where he completed some of the work on solid-propellant anti-aircraft rockets for the new Italian Navy. In 1953, Oberth returned to Feucht, Germany, to publish his book *Menschen im Weltraum (Man into Space),* in which he described his ideas for space-based reflecting telescopes, space stations, electric-powered spaceships, and space suits. After 1953, he moved to the United

---

[9]VfR began its first experiments with crude liquid propellant rockets in 1930. Their first rocket motor was called the Mirak, for Minimum Rocket. By the Spring of 1931, they tested their Repulsor rockets, which had two outside tanks. One was for the gasoline fuel and the other for the nitrogen pressurant which forced in the oxidizer into the combustion chamber. The liquid oxygen oxidizer tank was situated above the motor, though there were several varied designs. From August 1931, they began tests on a more simplified Repulsor IV model. In this configuration, the experimenters used the Congreve method of placing the tanks in line down in the center of the rocket. These types of rockets, some using liquid oxygen and alcohol as propellants, were tested over the space of a year and represented the VfR's last official rocket experiments.

States to again team up with von Braun on the development of rockets capable of reaching outer space. Their work led to the design and construction of the Saturn V rocket, which carried men to the Moon. He remained there for three years until retiring to West Germany in 1958.

1.4 P. Korolev and W. von Braun

During the 1920s and the 1930s, works by young scientists and astronautical engineers from Russia, and Germany turned Tsiolkovsky's theoretical works into hardware. The histories and biographies of these persons are well established. The following sections draw on these historical and biographical works[10] to pinpoint the contributions of these pioneers to the dawning of the space age, and the obstacles they faced in spreading their works into their economy and society.

1.4.1 The Soviet Union: Sergei Pavlovich Korolev

Sergei Pavlovich Korolev was born in the town of Zhitomir in Ukraine. He received vocational training in carpentry and in various academics at the Odessa Building Trades School, which gave its graduates a privilege of entering colleges without entrance exams. A 1913 air show inspired Korolev's interest in aeronautical engineering. After attending the Kiev Polytechnic Institute, he went on to the Moscow Higher Technical University in 1926. There he encountered the works of Tsiolkovsky[11] . At the time of his arrival in Moscow, there was a growing interest among the general public and engineers for space flights[12]. In 1927, the first world exhibition of Interplanetary Apparatus

---

[10]An extensive biography of von Braun can be found in Erik Bergaust (1976), while an authoritative biography of Korolev can be found in Encyclopedia Astronautica.

[11]It is less clear how Tsiolkovsky's writing influenced Sergei Korolev. In his 1934 book *Rocket Flight in the Stratosphere,* Korolev wrote "*He (Tsiolkovsky) founded the theory of rocket flight...and explored numerous issues related to manned flight at high altitude in outer space.*" According to Yaroslav Golovanov (1994), Korolev's biographer, the copies of Tsiolkovsky's books found in Korolev's personal library are covered in pencil notations.

[12]Russians have long had a spiritual fascination with space. For centuries, the people told parables, folk tales, and myths about space travel. A mystical early-20th century Russian philosophy known as Cosmism wanted humans to travel into the universe, recover the ashes of the deceased, resurrect the dead, and settle throughout the cosmos. Newspapers proclaimed that hundreds of star ships would soon venture out into the cosmos. Ordinary citizens organized competitions to build rockets to reach the edge of space. Between 1921 and 1932, Russian media published nearly 250 articles and more than 30 non fiction books about spaceflight.

and Devices was held in Moscow. Papers of Goddard, Oberth, and Max Valiers were on display.

It was in this innovative intellectual atmosphere that Korolev became interested in the possibilities of rocket-propelled aircraft. In September 1931, he devoted, together with F. Tsander[13], his free time to experiments carried out by the Moscow rocketry organization GIRD (Group for Investigation of Reactive Motion), one of the earliest state-sponsored centers for rocket development in the Soviet Union. There were two other institutions working on rocketry: the Leningrad GIRD and the Leningrad Gas Dynamics Laboratory (GDL). The GDL focused primarily on rocket-propelled engines, while Leningrad GIRD was involved with the engineering design, construction, and testing of the craft that would be powered using the engines that the GDL developed. Moscow based GIRD's engineers were working on the design of an engine using oxygen as an oxidizer, while GDL engineers were working on an engine using nitric acid and a hydrocarbon.

The GIRD lasted only two years before the military, seeing the potential of rockets, replaced it and the GDL with the RNII (Reaction Propulsion Scientific Research Institute), and Korolev was nominated deputy chief engineer. At RNII, a group of engineers, including Korolev, developed a series of rocket-propelled missiles and gliders during the 1930s. These efforts culminated in Korolev's RP-318, Russia's first rocket propelled aircraft powered by Glushko's ORM-65 rocket engine. The engine burned a mixture of kerosene and nitric acid. The leadership at RNII favored Glushko's nitric acid engines[14] over the alternative liquid oxygen propulsion systems developed by Mikhail Tikhonravov, Leonid Dushkin and Andrei Kostikov, all former members of the Mos-

---

[13]Friedrich Tsander was a Baltic-German rocket scientist. As a high school student in Riga, Tsander had been exposed to Tsiolkovsky's ideas and became fanatical about spaceflight, especially about traveling to Mars. Together with Korolev, he designed the first liquid-fueled rocket to be launched in the Soviet Union, GIRD-X, and made many important theoretical contributions to the road to space. He was the first, along with Konstantin Tsiolkovsky, to introduce the idea of solar sail as a mean for spacecraft propulsion. In 1925 Tsander presented a paper, "Problems of flight by jet propulsion: interplanetary flights," (see NASA (1964) in which he suggested that a spacecraft traveling between two planets could be accelerated at the beginning of its trajectory and decelerated at the end of its trajectory by using the gravity of the two planets' moons, a method known as gravity assist. He died prematurely in Moscow in 1933.

[14]Valentin Petrovich Glushko was a Soviet engineer and the principal Soviet designer of rocket engines during the Soviet/American Space Race. In the 1920s, he attended Leningrad State University where he studied physics and mathematics. From 1929-1930 he pursued rocket research at the Gas Dynamics Laboratory. He became a member of the GIRD (Group for the study of Rocket Propulsion Systems), founded in Leningrad in 1931.

cow-based GIRD. As a result, a whole group of Muscovite engineers, advocating liquid oxygen propulsion systems, felt sidelined by the top management. Sergei Korolev, who also came to RNII from GIRD, pragmatically sided with his top managers. This was a setback for the evolution of advanced propulsion systems in the Soviet Union.

Before a flight test of the rocket, Korolev and Glushko were thrown into the Soviet prison system in 1938, during the peak of Stalin's purges. Glushko was arrested on March 23. Korolev, denounced by Glushko, was arrested on June 7.

Figure 1.4 – Korolev's Booking photo 1938
Courtesy of Natalya Koroleva

The charges against Korolev and Glushko were *"subversion in a new field of technology"*. Glushko was sentenced to eight years in the Gulag. Korolev was sentenced to ten years hard labor in the Kolyma gold mines, a virtual death sentence in the most dreaded part of the Gulag. However, Stalin recognized the importance of aeronautical engineers in preparing for the impending war with Hitler. A system of sharashkas (prison design bureaus) was set up to exploit jailed talents. Glushko worked on various aircraft projects with other arrested scientists. Korolev was saved by the intervention of senior aircraft designer Sergei Tupolev, himself a prisoner, who requested his services in the TsKB-39 sharashka.

In September 1940, Korolev was transferred to work at Tupolev's bureau in Moscow. In 1942, after Tupolev's team was evacuated to Omsk, Korolev was transferred to TsKB-16 in Kazan where he served as Deputy Director for Flight Testing, although still officially a prisoner. There he was able to return to work on rocket engines for aircraft and missile propulsion. In November 1944, Korolev was given charge of a team of sixty engineers and required to provide a draft project for a Soviet counterpart to the V-2. He worked together with Glushko to design a two-stage rocket that used liquid oxygen/alcohol propellants and an autopilot for guidance. These designs evolved into the more refined RD-1 and RD-2 rockets with a seventy five kilometers range, for protection of the capital from high-altitude Luftwaffe attacks.

Korolev remained in prison until June 1944, when he, along with Tupolev, Glushko and others, was finally discharged by special government decree, although the charges against him were not dropped until 1957. Korolev was commissioned into the Red Army with the rank of colonel in 1945. He was among the experts flown to Germany to recover the technology of the German V-2 rocket. The Soviets placed a priority on reproducing lost documentation on the V-2, and studying its various parts and captured manufacturing facilities. That work continued in East Germany until late 1946, when 2,000 German scientists and engineers were sent to Russia through Operation Osoaviakhim. Most of the German experts, Helmut Groettrup being an exception, were engineers and technicians involved in wartime mass-production of V-2, and they had not worked directly with Wernher von Braun.

1.4.2 Germany: Wernher von Braun.

Wernher von Braun was born on March 23, 1912 in the small town of Wirsitz, in the Posen Province. He belonged to a noble family, his father bearing the German title of *Freiherr.* Beginning in 1925, von Braun attended a boarding school at Ettersburg Castle near Weimar. There he acquired a copy of *By Rocket into Planetary Space* by Hermann Oberth. Space travel fascinated von Braun, and from then on he applied himself to physics and mathematics to pursue his interest in rocket engineering. In 1930, he attended the Technische Hochschule Berlin, where he joined *the Spaceflight Society* (*Verein für Raumschiffahrt* or "VfR") and assisted Willy Ley in his liquid-fueled rocket motor tests in conjunction with Hermann Oberth. In spring 1932, he graduated from the *Technische Hochschule Berlin* (now Technical University of Berlin), with a diploma in mechanical engineering.

His early exposure to rocketry convinced him that the exploration of space would require far more than applications of the current engineering technology. Wanting to learn more about physics, chemistry, and astronomy, von Braun entered the Friedrich-Wilhelm University of Berlin for post-graduate studies and graduated with a doctorate in physics in 1934. While completing his graduate studies, he conducted in-depth research on rocketry, for which he received a grant from the Ordnance Department of Germany. The grant financed von Braun's research at a research station not far from Berlin. At the time, Germany was highly interested in Goddard's research. Before 1939, German scientists occasionally contacted Goddard directly with technical questions. Wernher von Braun used Goddard's plans[15] from various journals and incorporated them into the building of series of rockets that would culminate in the well known V-2.

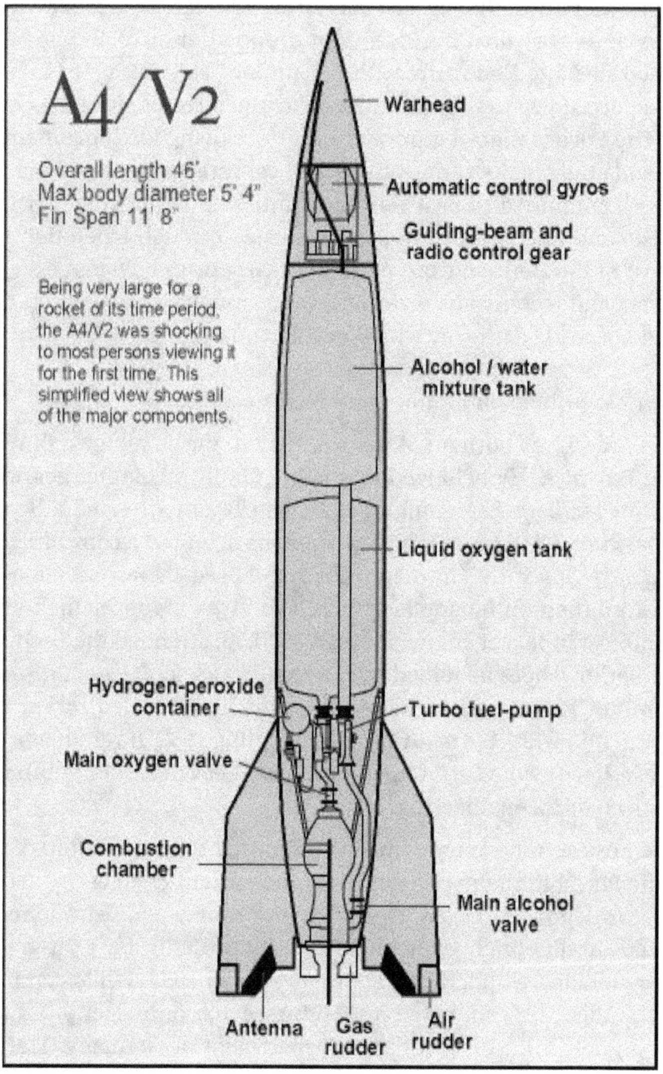

Figure 1.5 – V2
Source: http://imgarcade.com

---

[15]See NASA (1963).

After the collapse of the VfR, civilian rocket tests were forbidden by the new Nazi regime. Only military tests were allowed, and to this end, a larger facility was erected at the village of Peenemünde in northern Germany on the Baltic Sea. Moving to a new facility in the early 1940s, von Braun worked to again successfully launch rockets, as well as develop the supersonic anti-aircraft missile Wasserfall and the ballistic missile A-4. The A-4 became known as the V-2, meaning "Vengeance Weapon 2".

The V-2 was the most powerful rocket developed in this period. It used a seventy-four per cent ethanol/water mixture for fuel and liquid oxygen for oxidizer[16]. At launch the V-2 propelled itself for up to sixty five seconds on its own power. A program motor controlled the pitch to the specified angle at engine shutdown, after which the rocket continued on a ballistic free-fall trajectory. The rocket reached a height of 80 kilometers after shutting off the engine. The V-2 was guided by four external rudders on the tail fins, and four internal graphite vanes in the jet stream at the exit of the motor. The guidance system consisted of two free gyroscopes (a horizontal and a vertical) for lateral stabilization, and an accelerometer to control engine cut off at a specified velocity. The V-2 rocket laid the foundation for the liquid fuel missiles and space launchers used later on.

During this period, Von Braun had an ambivalent and complex relationship with the Nazi regime of the Third Reich. Was he a Nazi or was he a scientist? Most would like to believe the later. He officially applied for membership in the Nazi Party on November 12, 1937, and in 1940, he joined the SS and was given the rank of Untersturmführer. He was indirectly responsible in Peenemünde for the death of thousands of slave laborers. Later, von Braun would say that he accepted a position in the SS only out of necessity and that he only wore the uniform on one occasion. While his statement may be somewhat true, this does show that von Braun was willing to compromise too much to keep his research ongoing.

In the spring of 1945, the Soviet Army was about 160 kilometers from Peenemünde when von Braun and his staff decided to surrender to the Americans. The American high command was well aware of how important their catch was. von Braun had been at the top of *the Black List,* the code name for

---

[16]The fuel and oxidizer pumps were driven by a steam turbine. The steam was produced by concentrated hydrogen peroxide with sodium permanganate catalyst. The combustion burner reached a temperature of 2,500 to 2,700 °C . The alcohol-water fuel was pumped along the double wall of the main combustion burner. This regenerative cooling heated the fuel and cooled the combustion chamber. The fuel was then pumped into the main burner chamber through 1,224 nozzles, which assured the correct mixture of alcohol and oxygen at all times.

the list of German scientists and engineers targeted for immediate interrogation by US military experts. On June 20, 1945, the US Secretary of State approved the transfer of von Braun and his specialists to America. von Braun was among those scientists for whom the US Joint Intelligence Objectives Agency created false employment histories and expunged Nazi Party memberships and regime affiliations from the public record[17]. Once "bleached" of their Nazism, the US Government granted the scientists security clearance to work in the United States.

1.5 Obstacles to change

The first half of the 20th century was an age of change in many fields from science and technology, to politics, economics, society, and culture. In science, new areas of physics - special relativity, general relativity, and quantum mechanics - were developed. The internal structure of atoms was clearly understood. Theoretical works by Tsiolkovsky, Goddard, and Oberth demonstrated the possibility of escaping Earth gravity. Astronomy led to a much better understanding of the evolution of the universe. There were new and radical developments in life and human sciences, building on the progress made in the 19th century. New fields of mathematics studying the theory of computation were developed. The works of Alan Turing, including the Turing machine, and of John von Neumann were seminal.

One of the prominent traits of the first half of the 20th century was the dramatic growth of technology. But the growth of technology followed scientific discoveries of the second half of the 19th century[18]. Technological advances in the new fields of physics, mathematics, and astronautics were slow to materialize. This last section concentrates on astronautics and the obstacles to the development of the technology in this area. Advancement in astronautics

---

[17]Operation Paperclip was a secret program of the Joint Intelligence Objectives Agency (JIOA) in which more than 1,600 German scientists, engineers, and technicians, such as Wernher von Braun and his V-2 rocket team, were recruited in post-Nazi Germany and taken to the U.S. for government employment, primarily between 1945 and 1959; many were former members and some were former leaders of the Nazi Party. See A. *Jacobsen (2014).*

[18]Electricity availability powering factories and growing numbers and types of home appliances, and radio and telephone followed Maxwell scientific breakthrough on electromagnetism. The technology for the automobile existed already in the 19th century, but it took Henry Ford to make it accessible to the large public. Ford used the idea of the assembly line for automobile manufacturing. He paid his workers an unprecedented five dollars a day when most laborers were bringing home two dollars. This increased labor productivity. Higher earnings by workers induced many to purchase a new car.

and rocket engineering indicated the possibilities of accessing space for peaceful applications and/or to build powerful rockets for military purposes. The three countries, where major advancements in astronautics and rocket engineering were achieved, followed different paths in the development of these technologies because of their different cultural, institutional, and socio-economic environments.

1.5.1 The United States

In the United States, Goddard's efforts to develop rocket technologies were slowed down by the lack of interest by the US government and the private sector, the hostility of the press and of some academic circles, and the consequent lack of financing. At the end of the first world war, private concerns preoccupied most Americans. They turned inward and concentrated on domestic affairs. Three Republican presidents[19] occupied the White House during the 1920s. All of them were strongly supporting business, tried to reduce to a minimum government interventions in the economy, and were isolationists. It is no surprise that the request for government financial support by Goddard to develop rockets to carry instruments to study the atmosphere was rejected.

Due to the isolationist sentiment prevailing in the country in the 1920s, the initial efforts of Goddard to interest the military also failed. Private companies showed no interest in rocket technology. This was partly due to technological challenges, high risks, and long term return on investments, and extreme cost of hardware failure. But private companies' choices were also influenced by lack of interest from the government, as the selection of a technology is determined not only by conditions inside and outside the firms in the industry, but also by the characteristics and behavior of agents, such as governments, outside the industry.

In the 1930s, a new era commenced in American history. The federal government under Roosevelt experienced a vast expansion in its authority, initially over the economy. In the late 1930s, the aggressive policies of Germany and Japan, induced the federal government to create new institutions and to reinforce financially existing ones to cater the requirements of defense mobi-

---

[19]Warren Harding, a conservative, was swept into office by a landslide victory in 1920. He proved an inept president, and his administration was racked by scandals, including that of Teapot Dome. Calvin Coolidge, who succeeded to the office on Harding's death (1923), worshiped business as much as he detested government. Herbert Hoover, an engineer, brought to the presidency (1929-33), which to him represented the fullest expression of individualism.

lization. In 1938 and 1939, the National Advisory Committee for Aeronautics[20] (NACA) received additional funds for new aeronautical research laboratories in California and Ohio[21]. In 1940, President Roosevelt approved the creation of the National Defense Research Committee. The organization operated financially on a hand-to-mouth basis with monetary support from the president's emergency fund. It established a cooperation between civilian scientists and the military. On June 28, 1941, Roosevelt established the Office of Scientific Research and Development (OSRD) with the signing of Executive Order 8807. The OSRD was on a firmer financial footing than the NDRC since it received funding from Congress, and had the resources and the authority to develop technologies with or without the military.

These new institutions played a critical role in a crash program to create an atomic bomb[22]. The NACA contributed to overcoming the inferiority of American aircraft engines, and to the development, together with the British, of radar. However, there was a serious omission in NACA's portfolio: guided missiles. While the NACA had some success developing unguided rockets, it

---

[20]The National Advisory Committee for Aeronautics (NACA) was a US federal agency founded on March 3, 1915, to undertake, promote, and institutionalize aeronautical research. On October 1, 1958, the agency was dissolved, and its assets and personnel transferred to the newly created National Aeronautics and Space Administration (NASA).

[21]In December 1938, NACA asked for $11 million to establish a new aeronautical research laboratory in Sunnyvale, California, to supplement the existing Langley Memorial Aeronautical Laboratory. In 1939, funding for the new center, now known as the Ames Research Center, was finally approved. In the same year, NACA asked for funding to build a third center in Ohio, which became the Glenn Research Center.

[22]The discovery of nuclear fission by German chemists Otto Hahn and Fritz Strassmann in 1938, and its theoretical explanation by Lise Meitner and Otto Frisch, made the development of an atomic bomb a theoretical possibility. There were fears that a German atomic bomb project would develop, especially among scientists who were refugees from Nazi Germany and other fascist countries. In August 1939, Leó Szilárd and Eugene Wigner drafted the Einstein–Szilárd letter, which warned of the potential development of "extremely powerful bombs of a new type". The letter was delivered to Roosevelt. It urged the United States to take steps to acquire stockpiles of uranium ore and accelerate the research of Enrico Fermi and others into nuclear chain reactions. Roosevelt called on Lyman Briggs of the National Bureau of Standards to head the Advisory Committee on Uranium to investigate the issues raised by the letter. The Advisory Committee on Uranium became part of the National Defense Research Committee (NDRC) when that organization was formed on 27 June 1940. The NDRC Committee on Uranium became the S-1 Section of the OSRD; the word "uranium" was dropped for security reasons. On 9 October 1941, President Roosevelt approved the atomic program. Roosevelt chose the Army to run the project rather than the Navy, because the Army had more experience with management of large-scale construction projects. He also agreed to coordinate the effort with that of the British.

had nothing comparable to the V-1 and the V 2. This represented an entire field that had been left to the enemy. They did not seek the advice of Goddard.

1.5.2 The Soviet Union

In the Soviet Union, after the Bolshevik revolution of 1917 and the Russian civil War (1918–1920), a period called the new economic policy (1921–1927) allowed for a mixed economy to flourish. Books, pamphlets, and even some newspapers could be published independently from the state. A proliferation of popular books, newspaper articles, and pamphlets on air and spaceflight filled the popular press. Soviet readers became part of a cosmopolitan readership throughout Europe engaged in news on exploration of the cosmos. This continued a pre-revolutionary fascination with the stars, heavens, and the universe beyond. Astronomy and amateur space societies proliferated in Soviet Russia until the 1930s, and genuinely were generated from below. Within this economic and political context, air and spaceflight, along with astronomy, became popular themes in the mass media.

The Soviet aeronautical and rocketry craze was certainly part of a pan-European phenomenon, as the reporting of aeronautical feats in Europe were popular news items and were anticipated well ahead of time. Western technological developments were practically revered in the Soviet newspapers of the 1920s. Readers were exposed to news on global developments in aeronautics and rocketry. America itself was portrayed as a symbol and emblem of how technology was transforming modern culture. Soviet readers believed they were part of a cosmopolitan readership that could synthesize European, American, and Russian developments of rocketry and aeronautics in general.

Though interest in spaceflight had predated the 1917 Russian revolution, certain groups in the 1920s, such as the Biocosmists, believed in the importance of spreading ideas on interplanetary travel for public consumption. They included among their group the renowned geochemist and science popularizer V. Vernadskii and the space visionary Tsiolkovski. The Biocosmists were heavily influenced by the ideas and writings of the Russian pre-revolutionary philosopher Nikolai Fedorov. Fedorov had worried that Earth was overcrowded and believed that humans could overcome this Malthusian pressure by exploring and colonizing space.

During the 1920s, professional science educators also served as popularizers of space flight and rocketry. Those Russian intellectuals, such as the Leningrad physics professor Perel'man, had more didactic purposes in mind. Perel'man published many articles on rocket science and space travel in several widely distributed popular journals, such as *In Nature's Workshop*. These articles had an educational focus, attempting to explain the basics of gravitational forces and rudimentary rocketry to a popular audience. It is no surprise that the experimental rocketry works of engineers, such as Korolev, flourished during this period.

With the advent of Stalin in the 1930s, all this would change. During Stalin's tenure, Lenin's New Economic Policy was replaced with a centralized command economy, industrialization, and collectivization. By the mid-1930s, a cultural shift had occurred in Russia, dubbed by Nicholas Timasheff as "the great retreat"[23]. Timasheff, and some current cultural historians, have argued that, during the Stalinism period, Russia embodied a retreat away from a cosmopolitan culture back toward more nationalistic themes. Along with Stalin's Soviet nationalist cultural campaigns, it would begin a contest with the West for technological superiority that dumped the early popular cosmopolitan enthusiasm, and politicized and ultimately nationalized scientific and technological research. It was within this intellectual and institutional environment that the government shifted away from rocketry and space exploration, while the Soviet aeronautical feats were glorified and popularized through propagandist means by the Soviet press in preparation of the expected war with the West.

1.5.3 Germany

In Germany, the first world war[24] was the initial event that brought substantial changes in politics and economics. The First World War devastated Germany with millions of casualties and the loss of land and property. A slow process of economic recovery began following the war. While the economy of the United States flourished in the 1920s, the German economy took much longer to recover and grow. It was not until the mid-1920s that the German economy surpassed 1913 production level. Newer industries like automotive, radio, and commercial aviation spurred growth, as did the implementation of standardized mass-production in industry. In considering the general political situation at the beginning of 1920, the Liberal and Radical parties in combination with the so-called Majority Social Democratic Party had held power continuously, and had been strikingly confirmed in their position by the general election held in January 1919. Culturally, Berlin remained the leading center of physics, mathematics, and chemistry.

It was in this climate that space travel and rocketry gained popularity in German speaking Central Europe following the June 1923 publication of Oberth's book *Die Rakete zu den Planetenräumen* . The VfR, which had been formed in June, 1927, immediately began designing and building a variety of rockets. The leading members of the VfR established intense cultural ex-

---

[23]See S. M. Timasheff (1946).
[24]See A. Sommariva et al. (1987).

changes with the United States and Russia. This followed a general trend among German intellectuals toward a more liberal society after the horrors of the First World War. Militarism, as typified by Germany, was defeated and discredited. As Blinkhorn[25] argues, the liberal themes were ascendant in terms of cultural pluralism, religious and ethnic toleration, national self-determination, free-market economics, representative and responsible government, free trade, unionism, and the peaceful settlement of international disputes through a new body, the League of Nations.

The severe economic recession of 1929-1932 changed all this. The beginning of the economic recession in Germany can be attributed to the halting of capital inflows, particularly from the United States. The German economy was particularly hit by these developments because of the need to finance large current account deficits and war repayments. In order to restore the external balance, the German authorities were confronted with the alternatives of departing from the gold standard and of maintaining full employment[26] or of deflating their economy. Fears of inflation led to follow the second alternative. In 1930, the German government initiated a policy of balancing the budget through raising taxes and reducing government expenditures. These policies, together with the severe downturn of the international business cycle induced a fall in the net national product of about eleven per cent both in 1930 and in 1931. The rate of unemployment rose steadily from about 6 per cent in 1929 to 17 per cent in 1932.

Under the severe economic recession of 1929-1932, the social cement of the country crumbled, and the Nazi Party took power. Nazism subscribed to theories of racial hierarchy and Social Darwinism, identifying the Germans as a part of what the Nazis regarded as an Aryan or Nordic master race. The Nazis aimed to unite all Germans living in historically German territory, as well as gain additional lands for German expansion under the doctrine of Lebensraum, and exclude those who they deemed either community aliens or "inferior" races. The international and bottom up approach of the VfR and its ultimate goal of space flight were ended in 1934, because of Adolf Hitler's fears of the significant ties of the VfR to the international community and the interest of the Army in ballistic rather than space flight. The German Army understood the military applications of rockets and several VfR members, notably Wern-

---

[25]See M. Blinkhorn (2000).

[26]There were proposals for devaluation and credit expansions. In particular, William Lautenbach, an official of the Ministry of Economics, proposed a plan to expand bank credit to finance public works, which was rejected by the German government. These reformers were politically opposed to Hitler, but ironically supplied him with ideas that previous democratic governments refused to use.

her von Braun, went on to be important in German Army rocket program. After the end of the war, Germany was in crumbles and divided. For a long time, it disappeared from the leading group of nations pursuing space flight.

1.6 Lessons from history

The first half of the 20th century saw the first stage of the innovation process. Although theoretical works on astronautics were well advanced in the first half of the 20th century, advancements in astronautical engineering and in peaceful use of space were slow. Despite the visionary works of Tsiolkovsky, Goddard, and Oberth, none of the countries under analysis pursued the possibilities of accessing space for peaceful applications. The collapse of the international order and the dominant nationalism in his various expressions were some of the causes of the slow advancement of peaceful space applications. Germany pursued actively the development of powerful rockets for military purposes, due to a curious oversight in the Treaty of Versailles, which did not include rocketry in its list of weapons forbidden to Germany, and the German Army understanding of the military applications of rockets.

Despite the very slow advancements in peaceful space applications, the important legacy of this period is fourfold. The theoretical foundations of astronautics were well established. The works of Tsiolkovsky, Goddard, and Oberth and their vision of space exploration influenced several young people, among which von Braun and Korolev, who went on to do significant work on space programs later on. The creation of increasingly powerful rockets in Germany contributed to the revived interest both in the United States and the Soviet Union on rocketry after the war. This led to the construction of more powerful rockets for military purposes, which, with some modifications, allowed later on access to space for peaceful applications. Finally, under the pressures of the economic depression and the aggressive foreign policies of Germany and Japan, the United States introduced profound changes in its institutions that will play an important role later on in the development of space technologies and the space economy.

At the end of the Second World War, the international environment was profoundly changed. The United States and the Soviet Union emerged as the dominant countries. Western Europe and Japan were in crumbles. Under the pressures from the United States, countries outside the Soviet block reorganized themselves in democratic governments. Trade liberalization measures were introduced after years of isolation and abnormal international relations. But nationalism continued to pervade the Soviet Union policies. Countries within the Soviet block reorganized along the lines of collective and centralized economies. Soviet Union armies dislocated in those countries suppressed any tentative to divert from this path. The Soviet Union continued the technological contest with the West, and pursued aggressive foreign policies. Only the threat of a global nuclear holocaust prevented a clash of arms between the two super powers. The so called Cold War period began, which played a substantial role in the advancement of space technologies and the birth of the space economy.

Appendix - The Rocket Equation

The rocket equation is named after Russian scientist Konstantin Tsiolkovsky who independently derived it and published it in his 1903 work. The Tsiolkovsky rocket equation, or ideal rocket equation, describes the motion of vehicles that follow the basic principle of a rocket: a device that can apply acceleration to itself using thrust by expelling part of its mass with high velocity and thereby move due to the conservation of momentum. The equation relates the delta-v (the maximum change of velocity of the rocket if no other external forces act) with the effective exhaust velocity and the initial and final mass of a rocket, or other reaction engine[27].

For any such maneuver (or journey involving a sequence of such maneuvers):

$$\Delta v = v_e * \ln(m_0 / m_f) \quad (1)$$

where:

Δv = the maximum change of velocity of the vehicle (with no external forces acting).

m_0 = the initial total mass, including propellant.

m_f = final total mass without propellant, also known as dry mass.

v_e = the effective exhaust velocity.

ln = the natural logarithm function.

The rocket equation contains three variables. Given any two of these, the third becomes cast in stone. Hope, wishing, or tantrums cannot alter this result. Although a momentum balance, these variables can be cast as energies. They are the energy expenditure against gravity (often called delta V or the change in rocket velocity), the energy available in your rocket propellant (often called exhaust velocity or specific impulse), and the propellant mass fraction (how much propellant you need compared to the total rocket mass). The energy expenditure against gravity is specified by where you want to go. Next, we need to choose the type of rocket propellant, thus specifying the available energy. Currently, all our human rated rocket engines use chemical reactions (combustion of a fuel and oxidizer) to produce the energy. There are limits to the quantity of energy that can be extracted from chemistry and thus there exists bounds placed outside of human control on the energy one can pack into a rocket. With these two variables set, the rocket mass fraction is now dictated by the rocket equation. We must build our rocket within this mass fraction or it will not reach its destination.

---

[27]For a derivation of the rocket equation see J. Muzio et al. (1992)

The rocket equation captures the essentials of rocket flight physics in a single short equation. The rocket equation only accounts for the reaction force from the rocket engine; it does not include other forces that may act on a rocket, such as aerodynamic or gravitational forces. As such, when using it to calculate the propellant requirement for launch from (or powered descent to) a planet with an atmosphere, the effects of these forces must be included in the delta-v requirement. In what has been called "the tyranny of the rocket equation", there is a limit to the amount of payload that the rocket can carry, as higher amounts of propellant increment the overall weight, and thus also increase the fuel consumption.

The rocket equation can be applied to orbital maneuvers in order to determine how much propellant is needed to change to a particular new orbit, or to find the new orbit as the result of a particular propellant burn. When applying to orbital maneuvers, one assumes an impulsive maneuver, in which the propellant is discharged and delta-v applied instantaneously. This assumption is relatively accurate for short-duration burns such as for mid-course corrections and orbital insertion maneuvers. As the burn duration increases, the result is less accurate due to the effect of gravity on the vehicle over the duration of the maneuver. For low-thrust, long duration propulsion, such as electric propulsion, more complicated analysis based on the propagation of the spacecraft's state vector and the integration of thrust are used to predict orbital motion.

The rocket equation should be modified in case of a relativistic rocket. A relativistic rocket refers to any spacecraft that travels at a velocity close enough to light speed for relativistic effects to become significant. The meaning of "significant" is a matter of context, but generally speaking, a velocity of at least 50 per cent of the speed of light is required. Above this speed special relativity is required to accurately describe motion, whereas below this speed sufficient accuracy is provided by Newtonian physics and the Tsiolkovsky rocket equation. Relativistic rockets are usually seen discussed in the context of interstellar travel, since most would require a great deal of space to accelerate up to those velocities.

As with the classical rocket equation, one wants to calculate the final velocity change Δv that a rocket can achieve after expelling all its reaction mass and being reduced to a rest mass of m1 in the inertial frame of reference where the rocket started at rest depending on the specific impulse Isp. In order to make the calculations simpler, we assume that the acceleration is constant (in the rocket's reference frame) during the acceleration phase; however, the result is nonetheless valid if the acceleration varies, as long as Isp is constant. The relativistic rocket equation is the following[28]:

$$\Delta v = c * \tanh \left(\left(I_{sp}/c \right) * \ln \left(m_0 / m_1 \right) \right) \quad (2)$$

where c is the speed of light and tanh the tangent hyperbolic trigonometric function.

[28] For a derivation of the relativistic rocket equation see Robert L. Forward (1995)

Bibliography

Blinkhorn M.: "*Fascism and the Right in Europe, 1919-1945*", Harlow, Longman, 2000.

Emme E. M.: *"A History of Space Flight"*, ed. Holt, Rinehart and Winston, 1965.

R. Forward: *"A Transparent Derivation of the Relativistic Rocket Equation"*, paper presented at the 31 st AIAA/ASME/SAE/ASEE Jet Propulsion Conference and Exhibit, July 10-12,1995, San Diego, CA.

Goddard R. H.: *" Rockets: Comprising "A Method of Reaching Extreme Altitudes" and "Liquid-Propellant Rocket Development"*, published by the American Rocket Society, 1946.

Jacobson A.: *" Operation Paperclip: The Secret Intelligence Program to Bring Nazi Scientists to America", Little, Brown and Company, New York*

Moore W.:"*A Treatise on the Motion of Rockets. To which is added, An Essay on Naval Gunnery", ed. G. and S. Robinson, London (1813).*

Muzio J. et al.:"Celestial Mechanics and Dynamical Astronomy", Kluwer Academic Publishers, 1992.

NASA: *"von Braun's recollections of childhood/early experience in rocketry"*, http.history.mfsc.nasa,gov/vonbraun/recollections

Oberth H.: *"The Rocket into Planetary Space"*, ed. De Gruyter Oldenbourg, Berlin/Boston, 2014.

Oberth H.: *" The Ways of Space Flight"*, NASA TT-F-622, 1972.

Sommariva A. et al.: *"German Macroeconomic History. A Study of the Effects of Economic Policy on Inflation, Currency Depreciation, and Growth"*, ed. St. Martin's Press, 1987.

Timasheff N. S.: *"The Great Retreat: The Growth and Decline of Communism in Russia"*, ed. E. P. Dutton & Company, New York, 1946.

Tsander F. A.: *"Problems of flight by jet propulsion interplanetary flights"*, NASA, nasa_techdoc_19650001039, 1964.

Chapter 2

The dawn of the space economy, 1950-1970: the innovation process stage

"There is nothing so great or so crazy that a technological society can not be moved to do, provided that it is physically possible."

Freeman Dyson

The dawn of the space economy can be traced to the Cold War and the geopolitical competition between the United States and the Soviet Union. At the end of the Second World War, the United States and the Soviet Union emerged as the two superpowers with profound economic and political differences[1]. In the early 1950s, development of intercontinental ballistic missiles became a priority for both governments, which were primarily designed for nuclear weapons delivery. The two superpowers never engaged directly in full-scale armed combat, but they were heavily armed in preparation for a possible all-out nuclear world war. Each side had a nuclear deterrent that discouraged an attack by the other side, since such an attack would lead to total destruction of the attacker: the doctrine of mutually assured destruction (MAD). These heavy lifters and/or their modifications were instrumental for launching into low earth or geostationary orbits Earth observation and military communication satellites.

Aside from the build up of the two sides' nuclear arsenals, and the deployment of conventional military forces, the struggle for dominance was expressed via proxy wars around the globe, psychological warfare, massive

[1]The Soviet Union was a Marxist–Leninist state ruled by its Communist Party, who in turn were ruled by a dictator or a small committee ("Politburo"). The Party controlled the press, the military, the economy, and all organizations. It also controlled the other states in the Eastern bloc, and funded Communist parties around the world. In opposition stood the West, dominantly democratic and capitalist with a free press and independent organizations. The United States, being the main power within the Western block, began a strategy of global containment to challenge Soviet power, extending military and financial aid to the countries of Western Europe and creating the NATO alliance.

propaganda campaigns and espionage, rivalry at sports events, and technological competitions. Under pressure of geopolitical competition, the United States and the Soviet Union created vast national space programs in the 1960s, culminating in American astronauts landing on the Moon in 1969, and the the deployment in orbit of military satellites for telecommunication and Earth observation, mainly for reconnaissance and spy activities, by the United States and the Soviet Union.

2.1 Military programs

After the end of the war in Europe, rocket and missile development became a national priority in the Soviet Union. A new institute was created for that purpose in the suburbs of Moscow under the military control of Dmitriy Ustinov, with Korolev serving as a chief designer. Korolev's team oversaw German specialists - including Helmut Gröttrup and Fritz Karl Preikschat - captured in Peenemünde and Mittelwerk. With blueprints produced from disassembled V-2 rockets, the team began producing a working replica of the rocket. This was designated the R-1, and was first tested in October 1947. A total of eleven rockets were launched with five hitting the target. This was similar to the German hit ratio, and demonstrated the unreliability of the V-2.

In 1947, Korolev's group began working on more advanced designs. The R-2 doubled the range of the V-2, and was the first design to utilize a separate warhead. This was followed by the R-3, which had a range of 3,000 kilometers, and thus could reach England. The Soviets continued to utilize the Germans' expertise on V-2 technology for some time. But in 1950 the Ministry of Defense decided to dissolve the German team and repatriated the German engineers and their families.

Under Korolev leadership, the Soviet Union produced the R-7 Semyorka, the world's first true intercontinental ballistic missile. It was a two-stage rocket with a maximum payload of 5.4 tons, sufficient to carry the Soviets' bulky nuclear bomb, to an impressive distance of 7,000 kilometers. After several test failures, the R-7 successfully launched in August 1957 sending a dummy payload to the Kamchatka Peninsula. Korolev's consuming passion was the defense of his country and space exploration. Having achieved the first objective, he started to think about the second one. Korolev was keenly aware of the orbital possibilities of the rockets being designed as ICBMs. He conceptualized using the R-7 to launch a satellite as early as 1953, and suggested to the Academy of Sciences of the Soviet Union the concept for sending a dog into space. But the Communist Party was uninterested at that time.

The United States captured a large number of German rocket scientists, including von Braun, and brought them to the United States as part of Operation Paperclip. Under von Braun leadership, the V-2 evolved into the American Redstone rocket. The Redstone was designed as a surface-to-surface missile for the US Army[2]. It was the first large American ballistic missile and a direct descendant of the German V-2 rocket. Redstone was capable of flights from 93 to 323 kilometers. It consisted of a thrust unit for powered flight and a missile body for overall missile control and payload delivery on target[3]. The missile body separated from the thrust unit 20 to 30 seconds after the termination of powered flight, as determined by the preset range to the target. The body continued on a controlled ballistic trajectory to the target impact point. The thrust unit continued on its own uncontrolled ballistic trajectory, impacting short of the designated target.

The Redstone program proved to be a bone of contention between the Army and Air Force due to their different ideas of nuclear warfare. The Army favored using small warheads on mobile missiles as tactical battlefield weapons, while the Air Force, responsible for the ICBM program, wanted large cross-continental missiles that could strike Soviet targets and rapidly cripple the USSR's infrastructure and ability to wage war. The Air Force sponsored the design and construction of Atlas and Titan rockets. Atlas is a family of American missiles and space launch vehicles. The original Atlas missile was designed in the late 1950s and produced by the Convair Division of General Dynamics, to be used as an intercontinental ballistic missile[4]. Titan is a family of United States expendable rockets used between 1959 and 2005. Titan rockets were built by Martin Company (later Lockheed Martin Corporation) for the US Air Force in the late 1950s. Titans were part of the American intercontinental ballistic missile deterrent until the late 1980s.

---

[2]Redstone's prime contractor was the Chrysler Corporation. Rocketdyne Division of North American Aviation Company provided the rocket engines; Ford Instrument Company, division of Sperry Rand Corporation, produced the guidance and control systems; and Reynolds Metals Company fabricated fuselage assemblies as subcontractors to Chrysler.

[3]During powered flight, Redstone burned a fuel mixture of 25 percent water and 75 percent ethyl alcohol with liquid oxygen used as the oxidizer. Later Redstones used Hydyne, 60% unsymmetrical dimethylhydrazine and 40% diethylenetriamine, as the fuel. The missile body consisted of an aft unit containing the instrument compartment, and the warhead unit containing the payload compartment and the radar altimeter fuze.

[4]It was a liquid propellant rocket burning liquid oxygen and RP-1 fuel in three engines configured in an unusual "stage-and-a-half" or "Parallel Staging" design: its two outboard booster engines were jettisoned during ascent, while its center sustainer engine, propellant tanks and other structural elements were retained through orbital insertion (for orbital flights).

During this period, astronautical engineering made progress not only in the design of increasingly powerful rockets, but also in the solution of the problems posed by the return of a spacecraft into the atmosphere. While an orbital vehicle returning to the atmosphere had enough kinetic energy to vaporize itself due to friction with the air, it was known that meteorites could reach the Earth. The mystery was solved in 1951 when H. Julian Allen and AJ Eggers Jr. discovered that a beveled shape allows a more effective thermal shield. With this type of form, about 99 per cent of thermal energy is dispersed in the air rather than on the vehicle, allowing safe recovery of orbital vehicles. The discovery Allen and Eggers, initially treated as a military secret, was only published in late 1957[5]. This made possible the design of thermal shields adopted in later space missions, allowing astronauts to survive their re-entry to the Earth's atmosphere.

The early fifties were particularly frustrating for Von Braun because of the lack of interest of the United States government in space exploration. Von Braun developed a fascination for interplanetary flight since he was at school in Germany. During the war and in the late 1940s, he used his spare time to write a science fiction novel on a manned mission to Mars. He based his story on comprehensive engineering diagrams and calculations, included in the appendix to the manuscript[6]. The mission included seven passengers and three cargo ships. He planned a half-and-half year Mars' round trip.

The German space flight journal *Weltraumfahrt* published the appendix in a special edition in 1952. Later that year, Umschau Verlag in West Germany published the book as *Das Marsprojekt.* Henry J. White translated it into English. The University of Illinois Press published it under the title *The Mars Project* in 1953. In a series of articles in Colliers Weekly, von Braun developed the concept of a space station from which to launch spacecrafts for the Mars mission. This book and a series of articles on Colliers Weekly were his contribution to space exploration during this period.

---

[5]See National Advisory Committee for Aeronautics, Technical Note 4047, Washington DC, October 1957.

[6]The appendix to the *Mars Project* contained technical specifications for the mission to Mars. Von Braun envisioned a scientific expedition involving a fleet of ten spaceships with 70 crew members. The spacecrafts were to be assembled in Earth orbit using materials supplied by reusable space shuttles. He planned to use Hohmann trajectories to transport the spaceships from Earth orbit to Mars orbit. He calculated each ship's size and weight, the fuel they needed for the round trip, and each rocket's burn to effect the needed maneuvers.

Figure 2.1 – Design of von Braun space station in Colliers Weekly
Source: Chesley Bonestell Foundation.

2.2 The space programs

After Stalin's death in March 1953, he was succeeded by Nikita Khrushchev as First Secretary of the Central Committee of the Communist Party of the Soviet Union. In 1957, Korolev asked Khrushchev permission to use a modified R 7 rocket to launch a satellite (Sputnik 1) into orbit. He argued that the international prestige of the Soviet Union would rise by launching a satellite before the United States. In fact, the United States was discussing a possible satellite launch for the International Geophysical Year[7]. Although Khrushchev was not initially supportive, the spirit of Cold War competition was adequate to secure approval for the project.

[7]As early as 1950, a small group of scientists in the United States began discussing among themselves the possibility of using earth-circling satellites to obtain scientific information about the planet. In 1952, urged on by these same American scientists, the International Council of Scientific Unions (ICSU) proposed the International Geophysical Year (IGY), a cooperative scientific endeavor to study solar–terrestrial relations during a period of maximum solar activity. Some 67 nations agreed to conduct cooperative experiments to study solar–terrestrial relations during a period of maximum solar activity in 1957–1958. In July 1955, largely the same enclave of American scientists convinced Eisenhower that the United States should respond to the ICSU call for participation in the IGY by launching a scientific satellite. Eisenhower decision called for existing organizations within the Department of Defense to develop and launch a small scientific satellite "*under international auspices, such as the international Geophysical Year, in order to emphasize its peaceful purposes;...considerable prestige and psychological benefits will accrue to the nation which first is successful in launching a satellite . . . especially if the USSR were to be the first to establish a satellite.*"

The satellite was designed and constructed in less than a month. Sputnik 1 was a simple polished metal sphere no bigger than a beach ball, containing batteries powering a transmitter using 4 external communication antennae. Korolev personally managed the assembly at a hectic pace; and the satellite was placed in orbit in 1957 using a rocket that had been successfully launched only once before. International response to the accomplishment was electrifying, and political ramifications continued for decades. Nikita Khrushchev was pleased with this success, and encouraged the launch of a more sophisticated satellite in time for the 40th anniversary of the October Revolution on 3 November, 1957. This new Sputnik 2 had six times the mass of the Sputnik 1, and carried the dog Laika as a payload. The entire vehicle was designed from scratch within four weeks, with no time for testing or quality checks. It was successfully launched on 3 November 1957 and Laika was placed in orbit. There was no mechanism to bring the dog back to Earth, which died from heat exhaustion after five hours in space.

Korolev's planning for the piloted mission began in 1958 with design studies for the future Vostok spacecraft[8]. It was to hold a single passenger in a space suit, and be fully automated. The capsule had an escape mechanism for problems prior to launch, and a soft-landing and ejection system during the recovery. On 15 May 1960 an unmanned prototype performed 64 orbits of the Earth, but the re-entry maneuver failed. Four tests were then sent into orbit carrying dogs, and the last two were fully successful. After gaining approval from the government, a modified version of the R-7 was used to launch Yuri Alexeevich Gagarin into orbit on 12 April 1961. The first human in space and Earth orbit returned to Earth via a parachute after ejecting at an altitude of seven kilometers. Gagarin was followed by additional Vostok flights, culminating with 81 orbits completed by Vostok 5 and the launch of Valentina Tereshkova as the first woman cosmonaut in space aboard Vostok 6.

The successful launch of the Sputnik and of an orbiting capsule with an astronaut (Gagarin) created serious concerns in the United States. Although they did not create an immediate threat to the security of the United States, they indicated the advanced technological capabilities of the Soviet Union. In response to challenges posed by the Soviet Union space programs, the American military establishment proposed to take the lead in meeting these chal-

[8]The Vostok spacecraft was originally designed for use both as a camera platform (for the Soviet Union's first spy satellite program (Zenit), and as a manned spacecraft. This dual-use design was crucial in gaining Communist Party support for the program. The basic Vostok design has remained in use for some 40 years, gradually adapted for a range of other unmanned satellites. The descent module design was reused, in heavily modified form, by the Voskhod program.

lenges. But President Eisenhower, fearing the entry of the military-industrial complex in the space race, decided that *"space was to be used only for peaceful purposes."* He proposed the National Aeronautics and Space Administration (NASA) to Congress, which was promptly approved. NASA mission was to provide for research into the problems of flight within and outside the Earth's atmosphere and for other purposes. All existing space programs, including the Jet Propulsion Laboratory, were transferred to NASA. NASA thus became the titular owner of the United States space programs.

2.2.1 The Moon project

In May 1961, President Kennedy made one of the most inspired and visionary speeches[9] in the history of space exploration. In that speech, President Kennedy outlined a ten year program. The main targets were: landing men on the Moon (in the ticket was also a return trip to Earth), the deployment in orbit of satellites for telecommunication and earth observation, and the exploration of the solar system. Initially, Kennedy was aware of the large price tag of the Moon project, later renamed the Apollo program. Relations with the Soviet Union had also improved measurably since 1961. During a meeting with Khrushchev in Vienna, Kennedy suggested that the United States and the Soviet Union explore the Moon as a joint project. In the fall of 1963, Kennedy aggressively pursued a venture to turn the Apollo program into a joint effort[10]. But in both occasions, the Russian turned him down as they viewed the American offers as a ploy to open up Soviet society and compromise Soviet technology. Moreover, they thought that President Kennedy may not be reelected, and that a new President may well ditch the Moon project.

Initially, the Moon project captured the imagination of the general public and the interest of the law makers. Americans are quintessential progressives, generally accepting the notion that progress as a whole is good for humankind and that such progress typically occurs through advances in science and technology. Moreover, the United States was not yet fully involved in the war in Vietnam. There were no riots in the streets. The Moon project raised America's spirits in the face of the perceived Soviet space competition. America's status as a world superpower was largely based on a military apparatus that relied upon technology to project force and reduce risk. In this context, space was consistently represented as the new military "high ground." Vice president Lyndon B. Johnson embraced this point of view. But neither President Kennedy nor NASA knew how to achieve that goal. The staff at NASA was in a complete panic, but the pride inherent in fulfilling their president's commit-

---

[9] See www.archive.org/details/jfks19610525.

[10] See http://www.americanrhetoric.com/speeches/jfkunitednations1963.htm.

ment was a potent stimulus. Although Americans did not embrace von Braun because of his Nazi party's membership, he became the central figure of the Apollo program. Under his guidance, assisted by engineer Rocco Petrone[11], NASA developed the Saturn V rocket, a modified version of the Jupiter rocket. Von Braun later called it "an infant Saturn".

Saturn V became the cornerstone of the Apollo program, and it remains today the most powerful rocket ever built, and the only one to have carried humans beyond Earth orbit. The Saturn V's size and payload capacity dwarfed all other previous rockets which had successfully flown at that time. With the Apollo spacecraft on top, it stood 111 meters tall, and without fins. It was ten meters in diameter. Fully fueled, the Saturn V weighed 2,950 metric tons and had a low Earth orbit payload capacity originally estimated at 118 tons, but was designed to send at least 41 tons to the Moon. Later upgrades increased that capacity. During the final three Apollo lunar missions it deployed about 140 tons to LEO and sent up to 48.6 tons spacecraft to the Moon. The Saturn V was designed under the direction of Wernher von Braun and Arthur Rudolph at the Marshall Space Flight Center in Huntsville, with Boeing, North American Aviation, Douglas Aircraft Company, and IBM as the lead contractors.

At first, von Braun thought to assemble various modules (the Command/Service Module (CSM) and the Lunar Excursion Module (LEM)) in Earth orbit. Space shuttles would bring the modules into orbit. A second Saturn rocket would bring them to the Moon. But two Saturn V rockets increased costs and might cause delays in the mission. NASA engineer John Hubolt suggested the lunar orbit rendezvous (LOR) mission mode[12]. With LOR, a Saturn V rocket first would insert the modules into low Earth orbit and then propels them into lunar orbit using the last stage of the rocket. The lunar landing mission starts from lunar orbit. In this way, only one rocket would be used instead of two. A fierce debate convinced von Braun and NASA to adopt this solution in 1962. Undoubtedly, the decision to use LOR was vital for the manned landing on the Moon by the end of the decade, and to respect the

---

[11]Petrone graduated from the U.S. Military Academy at West Point in 1946. He earned a master's degree in mechanical engineering from the Massachusetts Institute of Technology in 1951. His career in rocket development began in the early 1950s in Huntsville, where he assisted in the to development of the Redstone rocket. He served as director of launch operations at NASA's Kennedy Space Center from July 1966 until September 1969 and afterwards as Apollo program director. His NASA colleagues described him as demanding. Clearly he played, along with von Braun, a vital role in the success of the Apollo missions.

[12]LOR was not new. First mention of LOR dates back to 1916. Yuri Kondratyuk, a Ukrainian engineer, proposed it (see Wilford J. (1969)). His analysis identified LOR as the most economical way of landing a man on the Moon.

commitment of the President. This solution saved time and money, but NASA paid a high cost. It discarded the assembly of the modules in low Earth orbit and the related infrastructure, which constituted the core of a space station. This had important consequences for longer term plans of human space exploration.

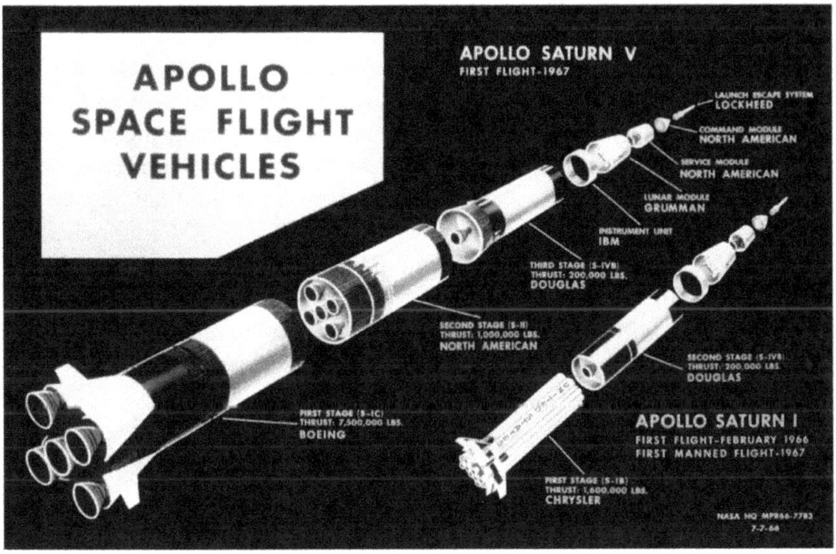

Figure 2.2 – Apollo Space Flight Vehicles
Source: http:// www.hq.nasa.gov

To meet the schedule imposed by Kennedy, NASA lowered the safety standards of the missions. This eventually caused headaches. During the first lunar landing mission, the on-board computer failed. With Armstrong manually piloting the landing module, he set it down on the Moon with only a few seconds of fuel remaining necessary to return to the earth re-entry vehicle. Even more serious was the case of the Apollo 13 mission, which took place a year later: an oxygen-tank explosion made the service module unusable during the cruise phase to the Moon. The astronauts instead occupied the lunar landing vehicle, uncomfortable but serviceable, and so returned safely to Earth. In retrospect, it was only with considerable good luck that, between July 1969 and December 1972, twelve American astronauts set foot on the Moon[13].

---

[13]In depth analysis of the Apollo program is in R. D. Launius (1994).

After Kennedy's assassination, President Johnson confirmed his support to the Moon project, but abandoned the idea of cooperating with the Soviets. In the wake of these events, the Soviet Union created its own program[14] for landing astronauts on the Moon. But the Soviet program suffered from three disadvantages. The first one was that all space programs were under the classified umbrella of the military, isolating the Soviet technological culture internationally. The second one was the slow development of the computer industry in the Soviet Union. A third disadvantage resulted from the division of work among several institutions (design bureaus). The design bureaus competed with each other, unlike the Apollo program which had one coordination center, NASA. But Korolyov's design bureau enjoyed a privileged position due to its successes in orbiting Sputnik and the first manned space capsule. Korolyov proceeded with two missions: cislunar and a manned landing on the Moon. The first program foresaw space capsule Soyuz 7K-L1 launched by the UR-500K rocket (Proton). A second one envisioned the same space capsule launched by N-1 rocket. The Soviet Union canceled the first program in 1970 and second one in 1974. The success of the Apollo mission and a series of launch failures of the N-1 rocket were the main reasons behind these cancellations.

Both programs enjoyed a high level of support among their respective populations and received large funding. Opinion polls conducted in the United States showed a consensus of over 60% of the population. Because of the closed nature of Soviet society, we have only anecdotal evidence on consensus among the Soviet population. But this evidence revealed deep and genuine support. The Apollo program cost twenty two billion current dollars (153 billion in 2016 dollars). It accounted for around 4.5 percent of the total expenditures of the federal government in 1967. While there are no official statistics of costs incurred by the Soviet program, American and British intelligence services offered a few estimates. They suggested figures close to those of the Apollo program.

These programs' economic benefits were large. In the short term, the multiplier effect stimulated the United States economy: for every dollar spent in the Apollo program, at least three dollars were fed into the economy[15]. Many inventions emerged as mere by-products of the Apollo missions, and the economic impacts of technological innovation predominated over the medium-long term. The financial support of the US government to research and development contributed to the rapid growth of the semiconductor and the computer industries. NASA scientists and engineers installed integrated cir-

---

[14]The Soviet space program history is told by I. Shklovsky (1991).
[15]See Chase Econometric Associates (1975); and H. R. Hertzfeld (1998).

cuits (IC) in their lunar and planetary spacecraft prior to the widespread use of these devices. At that time, NASA use of integrated circuits represented more than sixty per cent of total demand. Although there is no simple cause-effect relationship between space programs and the growth of the computer industry, it is true that the demand for these products by government space programs helped these industries' rapid advancement.

Chart 2.1 - NASA budget as per cent of federal expenditures

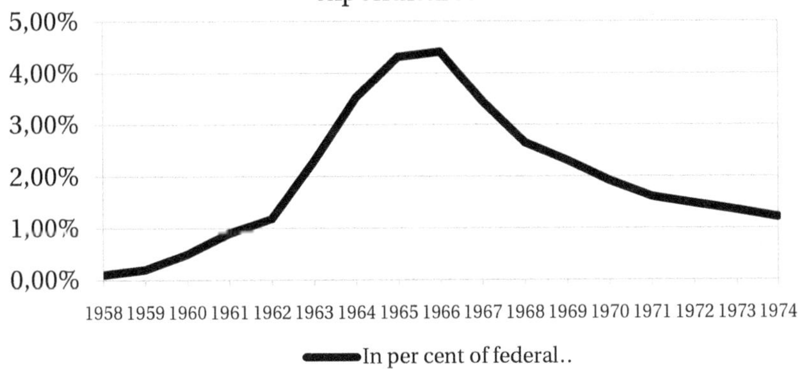

Credit: Author

Despite the obvious economic benefits, human space exploration came to a halt after the American astronauts' landings on the Moon. On September 15, 1969, NASA Task Group approved von Braun's Mars plan (see Appendix 2). In the following days, von Braun released and presented the final report to Congress, which did not approve it, although only by a few votes. There are several reasons behind the lack of support in Congress and the vanishing interest of the general public in large space programs. First, winning the technological competition with the Soviet Union was the main goal of the Apollo program. Once achieved, policy makers and opinion leaders raised tough questions: what were the long term goals of human space exploration and the benefits resulting from these missions? Since there were no compelling comprehensive answers, political support vanished.

Second, one of the project's weaknesses was the mission high costs. The program costs, excluding the costs of building the space station and the space shuttles, were estimated approximately at 50 billion current dollars (about 125 billion 2016 dollars) over thirteen years, with a peak funding rate of four billion dollars per year occurring between 1976 and 1978. Von Braun operated under the impression that whatever financial resources NASA might need, the government would provide them. But the United States government had other urgent priorities. Military expenditures had precedence because of the expanding Vietnam war, and of the Cold War reaching its peak.

Third, the Vietnam war and the following civil unrest caused the vanishing interest in large space programs among the general public, particularly among the young population. Families were starting to realize that both spouses needed to join the workforce to make ends meet. Racial equality was being confronted head on. Families were sacrificing their sons in an endless win-less war. They viewed large federal programs with distrust. It was these groups who protested the space program. Although they were a minority, they were very vocal and made an impact on later decisions to cut back funding to the space programs.

2.2.2 The government satellites' programs

In fall of 1945, Arthur C. Clarke wrote a short article in *Wireless World* describing the use of satellites in geosynchronous orbits high above the world's land masses to distribute television programs[16]. Although he was not the originator of the concept of geostationary satellites[17], his contribution was in the analysis of the technical feasibility of orbiting communication satellites and the prediction that the coverage they would provide would be cost-effective relative to the ground-based systems the stations. His article had little lasting effects in spite of Clarke's repeating the story in his 1951/52 *The Exploration of Space*[18]. The earliest satellites were made for the armed forces of the United States and Soviet Union military forces.

After the 1957 launch of Sputnik, many in the United States started to consider the benefits and prestige associated with satellite communications. Because of Congressional fears of "duplication," NASA confined itself to experiments with "mirrors" or "passive" communications satellites[19] (ECHO), while the Department of Defense was responsible for "repeater" or "active"

[16]See Arthur C. Clark (1945)

[17]The notion of a geostationary space station equipped with radio communication was published in 1928 by Herman Potočnik.

[18]See Arthur C. Clark (1951/52).

[19] Passive satellites only reflect the signal coming from the source, toward the direction of the receiver. With passive satellites, the reflected signal is not amplified at the satellite, and only a very small amount of the transmitted energy actually reaches the receiver. Since the satellite is so far above Earth, the radio signal is attenuated due to free-space path loss, so the signal received on Earth is very, very weak.

satellites[20] which amplify the received signal at the satellite and provide much higher quality communications.

Project Echo was the first passive communications satellite experiment. Each of the two American spacecraft, launched in 1960 and 1964, was a metallized balloon satellite acting as a passive reflector of microwave signals. Communication signals were bounced off them from one point on Earth to another. They were placed in low earth orbit where a signal could be sent to them, reflected by its surface, and returned to Earth. In the intentions of the developers, such satellites would be used for radio and television communication around the curvature of the earth, and as navigational aids that could be seen by ship and air plane radars night and day, clear weather or cloudy. But passive satellites were fraught with technical problems, and later on NASA abandoned passive communications systems in favor of active satellites. One unintended result of this program was the following: the large horn antenna at Holmdel constructed by Bell Labs for the Echo project was later used by Arno Penzias and Robert Woodrow Wilson for their Nobel Prize-winning discovery of the cosmic microwave background radiation.

The Defense Department was interested in satellites for communications between their various units stationed around the world. They contracted AT&T, Space Technology Laboratories, and Hughes Aircraft to do the job. Courier 1B was the world's first active repeater satellite. It was built by a division of Loral Space & Communications. Courier 1B was launched on 4 October 1960. Proposed by the U.S. Army Signal Corps, Courier used approximately 19,000 solar cells and was the first satellite to use nickel-cadmium storage batteries. It had an effective message transmission rate of 55,000 bits per second. The launch vehicles were a Thor-Able-Star, comprising a modified USAF Thor IRBM first stage and a USAF Able-Star upper stage with a restartable liquid engine.

On March 1955, the United States Air Force officially ordered the development of an advanced reconnaissance satellite to provide continuous surveillance of selected areas of the Earth in order to determine the status of a potential enemy's war-making capability, and to monitor the Earth's surface using sensors for detection of ballistic missiles. Any ballistic missile would be detected within seconds, which greatly reduced the chances of a surprise attack by an enemy. The first generation type (Corona) took photographs, then ejected canisters of photographic film which would descend to Earth. Corona capsules were retrieved in mid-air as they floated down on para-

[20]Active satellites amplify the received signal before re-transmitting it to the receiver on the ground.

chutes. The Corona program was produced and operated by the Central Intelligence Agency Directorate of Science & Technology. Later, the spacecraft had digital imaging systems and downloaded the images via encrypted radio links. The Coronas were launched by a Thor-Agena rocket[21], which used a Thor first stage and an Agena booster.

The Defense Support Program (DSP) is a program of the U.S. Air Force that operates the reconnaissance satellites. The Defense Support Program replaced the 1960s space-based infrared Missile Defense Alarm System (MiDAS). The first successful launch of MiDAS was 24 May 1960 and there were twelve launches before the DSP program replaced it in 1966. The first launch of a DSP satellite was on 6 November 1970 and since then it has become the mainstay of the United States ballistic missile early warning system.

In October 1957, two physicists at Johns Hopkins University decided to intercept microwaves coming from the Sputnik. They used a twenty Mhz receiver to capture microwaves. They transformed them into sound through an amplifier and calculated the satellite's speed by using the Doppler effect. This is possible due to the satellite first approaching the stationary receiver and then receding at a fixed frequency. By analyzing the Doppler shift slope, they calculated the satellite's position. As in any university network, the news spread. The laboratory's deputy director asked if, knowing the satellite orbit and speed, an object's position on the ground could be determined. The following week, they delivered theoretical calculations on this problem. But they did not know that the United States Navy was already interested in how to position its submarines for continuous navigation and had asked the deputy director to inquire about the problem. Five years later, the United States Navy had five satellites in orbit performing the job. Their research was the forerunner of the Global Positioning System (GPS).

A vast majority of satellites that the Soviet Union launched during the 1960s carried out military missions. It is, however, difficult to disentangle military and civilian satellites in the Soviet Union. Publicly the Soviet Union denied the very existence of the military space program. At the beginning of the 1960s, in order to provide a public "camouflage" for its expanding military space program, the Soviet government adopted a policy of assigning Kosmos names to all military satellites reaching orbit. In addition, any non-military payloads, whose mission was supposed to remain secret, were also lumped together with the Kosmos series. First of all, they included failed missions, since the Kremlin could not publicly admit any failure in the prestige-driven space program. As in the United States, the Soviet Union armed forces used

[21]Thor-Agena was a series of orbital launch vehicles. They are thus cousins of the more-famous Thor-Deltas, which founded the Delta rocket family.

these orbiting satellites to research battlefield information, locate known or missing troops, ensure communication, and take pictures. Also, they developed the equivalent of the Defense Support program for detection of ballistic missiles.

2.3 The dawn of commercial satellite activities

The first person to carefully evaluate the various technical options in satellite communications and their financial prospects was John R. Pierce. In a 1954 speech and 1955 article[22], he elaborated on the utility of a communications passive satellite in low earth orbit, a medium-orbit passive satellite, and a 24-hour-orbit (geosynchronous) active satellite. He compared the communications capacity of a geosynchronous satellite, which he estimated at 1,000 simultaneous telephone calls, with the one of the first transatlantic telephone cable, which could carry 36 simultaneous telephone calls at a cost of 30-50 million dollars. Pierce wondered if a satellite would be worth a billion dollars.

By the middle of 1961, NASA had awarded a competitive contract to RCA[23] to build a medium-orbit (6,437 kilometers high) active communication satellite. NASA also awarded a sole source contract to Hughes Aircraft Company to build a 24-hour (geosynchronous) satellite. From that moment on, private companies entered into a potential profitable market outlined by Pierce and capitalized on the vast research and development programs led by NASA and the military. Bell Telephone Laboratories assisted in the NASA Echo 1 project. Knowledge gained working on Echo 1 helped Bell to develop Telstar, an experimental satellite that relayed television signals. AT&T built Telestar that NASA launched in 1962 on a cost-reimbursable basis, providing also some tracking and telemetry support. In the six months following the launch, stations in the United States, Britain and France conducted about 400 transmissions with multi channel telephone, telegraph, facsimile, and television signals.

By 1964, four satellites, two by each company, operated successfully in space. In 1964, the Communications Satellite Corporation (Comsat) was formed as a result of the Communications Satellite Act of 1962. The primary goal of Comsat was to serve as a public, federally funded corporation intended to develop a commercial and international satellite communication system. Although the corporation was government regulated, it was equally owned by some major communications corporations and independent investors. Comsat's initial capitalization of 200 million dollars was considered sufficient to build a system of dozens of medium-orbit satellites.

---

[22]See John R. Pierce (1955)

[23]The RCA Corporation was a major American electronics company, which was founded as the Radio Corporation of America in 1919. At its height as an independent company, RCA was the dominant communications firm in the United States.

In 1965, an international global satellite consortium (Intelsat) was formed. It was an intergovernmental consortium[24], which, in time, owned and managed a constellation of communications satellites providing international broadcast services. On 6 April 1965, Intelsat's first satellite, nicknamed Early Bird, was placed in geostationary orbit above the Atlantic Ocean by a Delta D rocket. It was designed and built by Hughes Aircraft Company. Early Bird was the first operational commercial satellite providing regular telecommunications and broadcasting services between North America and Europe. Near complete Earth coverage (excluding polar areas) was achieved with the launching of satellites into geosynchronous Earth orbit over the Pacific (1967), and Indian oceans (1969). A combination of more than 130 governments and international organizations control Intelsat. Intelsat, along with Inmarsat, which is used in international shipping, is open to use by all nations. The Intelsat consortium owns the satellites, but each nation owns their own earth stations. Global satellite commercial services era had begun.

In the Soviet Union, after the successful launch of the Sputnik, the Soviet Union continued to develop satellite technology with the Molniya series of satellites, which were launched in a highly elliptical orbit to enable them to reach the far northern regions of the country. To broadcast to these latitudes from a geostationary orbit (above the Earth's equator) would require considerable power due to the low elevation angles. A satellite in a Molniya orbit is better suited to communications in these regions because it looks directly down on them. The primary use of the Molniya's orbit was for the communications satellite series, primarily for military use. The same orbits, with slight adjustments, were also used by some Soviet spy satellites, with the apogee point over the continental United States. Although geostationary orbits are useful for observing the continental United States, Soviet sensor technology sometimes required high-contrast observing angles which could only be achieved from higher latitudes.

The first satellite in this series, Molniya 1, was launched on April 23, 1965. By 1967 six Molniya satellites provided coverage throughout the Soviet Union. During the 50th anniversary of the Soviet Union on October 1, 1967, the annual parade in Red Square was broadcast nationwide via the Molniya satellite network. In 1971 the Intersputnik International Organization of Space Communications was formed by the Soviet Union along with a group of eight socialist states (Poland, Czechoslovakia, East Germany, Hungary, Romania,

---

[24]In 1973, the name was changed into Intelsat, and there were 80 signatories. Intelsat provides service to over 600 Earth stations in more than 149 countries, territories and dependencies. By 2001, Intelsat had over 100 members. It was also this year that Intelsat was privatized.

Bulgaria, Mongolia, and Cuba). Its objective was the common use of communications satellites. It was created as the Eastern Bloc's response to the Western Intelsat organization.

Although the United States and the Soviet Union proceeded separately in the development of the space economy due to their political and ideological differences, an incipient international collaboration was initiated in 1963. Like radio frequency spectrum, the geostationary orbit around Earth is a limited natural resource. Both need to be shared fairly and in a way that avoids interference. Approximately fifty space vehicles have been launched in the course of the satellite and lunar and space probe programs of the Soviet Union and the United States. Radio frequencies were sometimes " cleared " on an informal basis for use in space programs. On other occasions, there was no attempt at "clearance." In 1963, the International Telecommunication Union[25] (ITU) held an Extraordinary Administrative Conference for space communications, which allocated frequencies to the various services.

2.4 The political economy of the space age during the 1950s and the 1960s

During the period under analysis, the space economy was dominated by the public sector through civil and military programs. Only two countries operated large space programs: the United States and the Soviet Union. Government expenditures were motivated by the external and internal socio-economic environment in which those governments operated. Defense considerations and geopolitical competition were the main external factors behind governments investments in the development of powerful rockets and satellite technologies. After the United States won the race to the Moon, NASA programs of human space exploration were canceled. NASA budgets were reduced with respect to the previous period. Its activities concentrated mainly on robotic space exploration of the solar system and Earth observation by satellite, motivated mainly by scientific advancements. Also the Soviet Union scaled down its space programs, but remained active on the construction of a space station.

Internal socio-economic environments differed substantially between the two countries. In the United States, the civilian nature of NASA made it possible for the American researchers to collaborate and disseminate scientific advances. Through government spending, in the form of government research and development programs on satellite technologies, and contracts,

---

[25]The International Telecommunication Union (ITU): it coordinates standards for telecommunications. The standardization efforts of ITU commenced in 1865 with the formation of the International Telegraph Union (ITU). ITU became a specialized agency of the United Nations in 1947.

private companies took part in these activities. As a result, public spending paid for research and development. Private companies, participating in government R &D programs, were then able to choose in terms of a probability distribution of different new satellite technologies.

Because of the environment private companies were operating in, the risks of hardware failures were reduced sensibly by the space activities of various government agencies. This was particularly true in the case of launching vehicles. Government experimentation with launching vehicles were characterized by many failures, especially in the early period. Governments were able to sustain these failure rates because they were motivated by strategic reasons and were not constrained by profit motivations. Once launching vehicles were well advanced and their failure rate declined, the business risk for future private business activities in space declined. In summary, because of the environment in which private space companies evolve (technological challenges, the extreme cost of hardware failure, and the premium placed on a successful performance record), civil and military space programs were instrumental to the early growth of commercial space activities. This period marked the second stage of the innovation process encompassing the development of new ideas into marketable products and processes.

In the Soviet Union, the actual work and industrial efforts for the space programs were instead run under the classified umbrella of the Ministry of General Machine Building, with its enormous and rapidly expanding network of design bureaus and production facilities. The military was its principal client. The military also owned and operated every launch site and the network of ground control centers. The ministry had to report to the Communist Party's Central Committee and the Commission on Military-Industrial Issues of the Council of Ministers. Work beyond defense contracts was given secondary priority. As a result of this critical dependence on the military, the Soviet aerospace industry relied entirely on domestic hardware, all the way down to the tiniest individual micro-components. This resulted in an internationally isolated technological culture that created enormous barriers to the development of their space programs, and ultimately led to the failure of the Soviet Moon program.

Despite the geopolitical competition between the United States and the Soviet Union, the dawn of the space economy required international cooperation. Uncertainty about property rights and the uses of resources could hinder the funding of private space activities by increasing the risk premium. Private companies and financial institutions were not likely to risk their capital and the considerable effort to develop telecommunication satellites, if significant legal or political difficulties can disrupt the supply of these assets. Prior to 1963, the astronautical services were not officially defined in any Convention,

Radio Regulations or officially by any government. The ITU Extraordinary Administrative Conference put an end to era of informal agreements between the United States and the Soviet Union. This was a fundamental contribution to the evolution of peaceful space activities. Although at that time only two countries had communication satellites in orbit, the ITU agreement was a worldwide one requiring international co-operation, and in time many nations benefited by this agreement.

International cooperation was manifested not only in the ITU agreement but also in other areas. The United States proceeded with its multinational initiative under the umbrella of the United Nations to develop a legal framework for peaceful space activities. This eventually led to the Outer Space Treaty and creation of the United Nations Committee on the Peaceful Uses of Outer Space. Under the terms of the Outer Space Treaty, space, including the Moon and other celestial bodies, cannot be appropriated through claims of sovereignty by any nation, thereby establishing the legal framework for peaceful space activities. In the scientific community, the role of an international space science union was assumed by the Committee on Space Research formed in 1958. Its unusual charter gave a mandate to both superpowers to appoint vice presidents. This arrangement opened an opportunity for dialogue and informal contacts between the United States and other nations interested in space activities. It allowed the United States to maintain the lead in space technology, but it was instrumental to the growth of space activities by many other countries after 1970.

Appendix - Von Braun Mars mission: the missed opportunity

In his presentation to Congress[26], von Braun outlined the mission's goals:

> "With the recent accomplishment of the manned lunar landing, the next frontier is manned exploration of the planets. Perhaps the most significant scientific question is the possibility of extraterrestrial life in our solar system. Manned planetary flight provides the opportunity to resolve this universal question thus capturing international interest and cooperation.......
>
> The 1981 manned Mars mission (1982 landing on Mars) is shown as an integral part of the total space program for the next two decades. The systems and experience resulting from the Apollo program and the missions proposed for the 1970's provide the technical and programmatic foundation for this undertaking. A 1982 manned Mars landing is a logical focus for the programs of the next decade. Although the undertaking of this mission will be a great national challenge, it represents no greater challenge than the commitment made in 1961 to land a man on the moon....
>
> In addition to observations of Mars, many other experiments and measurements will be made on both the Earth-to-Mars and Mars-to-Earth legs of the trip that are of prime scientific importance. The spacecraft represents a manned laboratory in space, free of the disturbing influences of the Earth. The fact that there will be two observation points, Earth and the spacecraft, permits several possible experiments regarding the temporal and spatial features of the interplanetary environment. In addition, the spacecraft can be used to supplement and extend numerous observations conducted from Earth orbital space stations, particularly in the field of astronomy."

During the 1960s, von Braun worked out a plan, within NASA, for human exploration of Mars. Initially, regular Saturn V launches would occur all through the early to mid-70s, building a space station and generally preparing the necessary infrastructure to be assembled in orbit the manned spaceships for the Mars mission. Von Braun tweaked his original Mars-expedition scenario in his book Mars Project to halve the manned spaceship's size. He contemplated the launch two nuclear-powered manned spaceships. This would enable the two spaceships to travel in convoy, providing back-up and mutual

[26]See Library of Congress (2010)

support. The two nuclear powered spaceships were planned to be reusable for future expeditions, thus reducing the long term investment costs of the project. He contemplated the launch of robotic vehicles powered by chemical propulsion to be sent in advance to Mars. They contained equipment and propellant needed for the return to Earth. On the return trip, some of these vehicles would transport samples collected during the mission. The remaining would be redirected to enter Venus' atmosphere for further research.

The following elements characterized the manned spaceships:

a) Each spaceship would consist of three Nuclear Shuttles strapped in tandem. The two side Shuttles would get the center one and its payload up to speed, then peel loose and re-enter Earth orbit for reuse, while the diminished mission craft would carry on. Two lateral modules fueled for velocity's changes to set the spaceships toward Mars. They separated and maneuvered back to rendezvous and docking with the space station. There, they refueled. They could then be reused for earth-moon shuttle services or future Mars expeditions.

b) The one central module powered by a nuclear engine developed in the NERVA project would provide the thrust for insertion: into Mars orbit; for the return trip to Earth; and necessary maneuvers to enter Earth orbit. Significant fuel savings resulted due to insertion in a highly elliptical orbit around Mars.

c) A habitable module for six astronauts in each spaceship. Each had resources for twelve astronauts, if one spaceship needed to be abandoned during the mission.

d) One expedition module in each spaceship. The module was derived from the cone-shaped Apollo Command Module. The modules would descend to Mars from highly elliptical orbit. They could accommodate six astronauts for a 60-day stay on Mars. Since the spaceships flew in convoy, two modules could descend onto Mars next to each other. The six astronauts would subsequently use the upper stage of the modules to rejoin the orbiting spaceships at the end of their Mars mission.

Von Braun contemplated two types of missions: one with a short duration and the other one with a long duration. These missions are called the conjunction and opposition classes[27]. Conjunction-class trajectories are typified by low-energy transfers and by a long stay at Mars, around 500 days. The long stay provides the time required by Mars and Earth to reach relative positions to make possible the minimum energy transfer for a return. Opposition-class trajectories are characterized by one low-energy and one high-energy return transfers, and by around 40 days stay at Mars. Since an opposition-class expedition requires more energy, it demands more propellant. In fact, it needs ten times the propellant mass than does a conjunction-class trajectory. A final decision was made on an opposition-class trajectory.

The proposed mission profile with a short stay on Mars was the following:

- November 12, 1981: Insertion in trajectory to Mars. Each spaceship's mass at the start was to be 727 tons. At the detachment of the two lateral modules, the mass of each spaceship would be reduced to 614 tons.
- August 9, 1982: insertion into the Mars orbit. The mass of each spaceship before insertion into Mars orbit was estimated at 295 tons and, after a successful insertion, 226 tons.
- August 10, 1982: the excursion modules disengage from the spaceship and head to the surface of Mars. Meanwhile, the astronauts remaining on board the orbiting spaceship would analyse the surface of Mars and its two moons.
- October 28, 1982: the manned spaceships insert into their return trajectory to Earth. The mass of each spaceship at the start is 172 tons.
- February 28, 1983: Venus flyby. This reduces the velocity of the spaceships. It enables the launch of the remaining robotic spaceships into the atmosphere of Venus to do scientific experiments.
- August 14, 1983: entry into Earth orbit and docking with the space station. The final mass of each spaceship is 72.6 tons.

With hindsight, von Braun's program was too ambitious because of the state of nuclear propulsion's technologies at that time. Another element not analysed by von Braun was how to protect astronauts from cosmic radiations and solar flares, which could endanger the astronauts' lives and sensible components of the spaceships. Also, descent/ascent vehicles through Mars atmos-

---

[27]The terms conjunction class and opposition class refer to Mars' position relative to Earth. In the former, Mars moves behind the Sun as seen from Earth halfway through the expedition. In the latter, Mars is opposite the Sun in Earth's skies at the expedition's halfway point.

phere were not tested on the ground, which could have caused the abortion of the mission. Von Braun envisaged that these problems could be solved between 1969 and the time of the launch of the spaceships to Mars. Also his costs' estimates were most likely underestimated. Including the costs of building a space station and the space shuttles, the actual investment costs would have been higher.

One will never know whether von Braun approach was too optimistic. But his plan remains a cornerstone project for solar system's human exploration. Technically, it established two important conditions for interplanetary missions: first, the use of nuclear-powered spaceships to reduce travel time and diminish astronauts' and sensitive spacecraft systems to cosmic radiation; and second, a space-port to assemble nuclear powered spaceships and to insert them on trajectories to target destinations. Although it is possible to conceive launching nuclear-powered spaceships from the Earth surface, the former solution reduces security risks associated with the second one.

Bibliography

Chase Econometric Associates: *"The Economic Impact of NASA R&D Spending: Preliminary Executive Summary."* NASA Technical Note 2741, April 1975.

Clark A. C.: *"Extra-Terrestrial Relays Can Rocket Stations Give World-wide Radio Coverage?".* Wireless World. October 1945.

Clark A. C.: *"The Exploration of Space",* paperback, ed. Kessinger Publishing, 2007.

Hertzfeld H. R.: *"The impact of NASA research and development expenditures on technological innovation and the economy",* 1998.

Launius R. D.: *"NASA: A History of the US Civil Space Program",* Krieger Publishing Company, 1994.

Library of Congress: *"Wernher Von Braun Papers",* Manuscript Division, Library of Congress, Washington, D.C., 1978, Revised 2010.

National Advisory Committee for Aeronautics: *" Technical Note 4047",* NACA, TN-4047, 1957.

Pierce J. R.: *"Orbital Radio Relays",* Journal of Jet Propulsion, 1955.

Potočnik H.: *"The problem of Space Travel: The Rocket Motor",* The NASA Historical Series, NASA SP-4026, 1995.

Shklovsky I.: *"Five Billion Vodka Bottles to the Moon",* Norton & Co, New York, NY, 1991.

von Braun W.: *"The Mars Project",* ed. The University of Illinois Press, 1953.

Wilford J.: *"We Reach the Moon; the New York Times Story of Man's Greatest Adventure",* Bantam Paperbacks, New York, 1969.

Chapter 3

The space economy between 1970 and today: the diffusion stage

"It is our task, both in science and in society at large, to prove the conventional wisdom wrong and to make our unpredictable dreams come true."

Freeman Dyson

Since its inception in the early 1970s, the space economy has grown rapidly. It is a mixed economy, comprising the commercial space sector and government space activities. The commercial space sector is a high-technology niche with a complex ecosystem. It has distinguishing features such as the use of cutting edge technologies and longer terms for both projects' development and return on investments. This chapter analyses the growth of the space economy, the institutions and policies that were put in place to regulate relationships among different agents participating in the national and global space economy, and its impact on the world economy. It gives emphasis particularly to the international aspect, as individual countries initiatives, conducted independently at the national level, risk to be harmful and to produce serious international frictions.

3.1 The commercial space sector

The commercial space industry encompasses: the *commercial space services,* including satellite broadcasting, telecommunication, Earth observation, geolocation, and global navigation equipment and services, and consumer services outside the space community, which rely on some satellite capacity for part of their revenues; and commercial *space infrastructure and support industries,* including satellite and rocket manufacturing, launch services, ground stations and related equipment.

The global commercial space industry has grown steadily since 2006 at an average annual growth rate of ten per cent. In 2014, it reached about 281 billion dollars. The turnover of the commercial space services represents about sixty seven per cent of total revenues of the sector, while the commercial space infrastructure and support industries' turnover represents about thirty three per cent of the total.

Chart 3.1 - Turnover of the commercial space industry

(in billion of dollars)

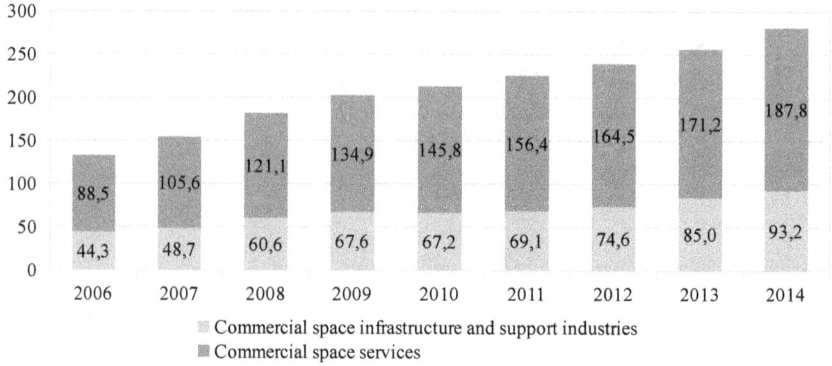

Credit: author. **Source data:** OECD (2014)

3.1.1 The commercial space service sector

During the early 1970s, the United States and Russia dominated satellite communications. Until 1970, long-distance communications in the United States were government-protected monopolies - the Bell System for telephones, and the three broadcasting networks (ABC, CBS, and NBC) for television and radio. In 1970, radical changes took place in the United States. New technologies were cropping up, such as cable television, mobile cellular telephones, microwave transmission, but were treated as appendages to the old system. Whitehead, a young staffer in the Nixon White House, saw things differently. He realized that the new technologies could supplant rather than supplement the status quo.

He proposed an "Open Sky" policy. Whitehead's simple idea was that any firm with the technical and financial wherewithal should be permitted to launch and operate its own communications satellite. The proposition was entirely permissive. It involved no central planning; no pilot projects; no restrictions, taxes, subsidies, or guarantees; and no requirements that anyone does anything. Initially, everyone of importance opposed it. But, precisely because the idea was permissive and hypothetical, opponents had a hard time writing up talking points against it. It was not illegal, international satellite communications were then a government monopoly, but the statute in question was silent on domestic satellites. The economies of scale that justified the Bell System monopoly seemed not to apply to satellites, so entry and price controls would be unnecessary. If a firm mounted a satellite that failed technically or commercially, the financial losses would be all its own.

The White House Office of Telecommunications Policy (OTP) was established in 1970. The OTP worked to unfreeze and deregulate the cable industry and implement the "Open Skies" policy. Open Skies rejected the monopoly industry structure that applied in international communications. OTP took a strong interest in supporting cable industry, because Whitehead saw that cable promises a revolutionary diversity and a fundamentally new system of communication. In summary, OTP deregulated the American market allowing entry of many companies, and launched a commercial space race. By the early 1980s, several firms, such as RCA and Hughes (as well as Bell), were operating satellites, leasing transmission channels to the broadcast networks and also to cable upstarts such as HBO, Turner, and CNN. Cable TV had sealed its arrival as an independent force by establishing its own public-service station, C-SPAN. AT&T lost its market power even before the 1984 antitrust break-up of its landlines-telecommunications monopoly. Competition and innovation produced low costs, high quality, endless variety of services, and a sheer ubiquity that would have been inconceivable before the 1970s.

Globally, an increasing number of countries entered the market in the 1970s, helped by the growing number of space agencies who invested in technological innovation. In the late sixties and the seventies, Europe and Japan created their own space agencies, followed by India and China, and other nations. By 2000, more than fifty countries have launched their own satellites, and seven countries have operational launchers. China and India rose as established space powers alongside the re-emergent Russian Federation. Asian countries led by China are outdistancing Europe in terms of number of launches and satellites sent in orbit. Geostationary Orbit (GEO) satellites are the backbone of the commercial satellite communications industry. Radio waves travel in straight lines at the microwave frequencies used for wideband communications, so a repeater is needed to send signals over long distances. Repeaters can be located on satellites. They can link places on the Earth that are thousands of kilometers apart, and large GEO satellites can serve one-third of the Earth's surface.

Satellite technology has now been extended to various services and applications, including products and services of information technology, television, GPS systems, and mobile phones. Direct Broadcast Satellite (DBS) television was developed in the United States to overcome "blind spots" in the coverage area of traditional terrestrial systems. Today, a DBS can send digital video signals from geostationary orbit directly to individual subscribers via small receiving dishes. There are other services that are focused on specific applications. These include radio-navigation satellite and meteorological-satellite services. Satellite technology's use in navigation, communications, meteorology and Earth observation gave rise to a growing stream of applications. These applications include air traffic control, transport, natural-resource management, agriculture, environmental and climate-change monitoring, and entertainment. Mobile satellite operators have traditionally provided communications to the narrower but profitable aeronautical and maritime markets.

Satellite technologies created new downstream uses and new markets. They include direct-to-home satellite television services providers, satnav consumer equipment and value-added services, and very-small apertures terminals providers, such as data handling and banking. Direct-to-home satellite television broadcast is almost universally available in OECD and emerging countries economies via one or more services, where the signal is received by satellite dishes and set-top boxes. Large media groups are providing the contents and actual satellite broadcasting, broadband and telephone services to everyday consumers. Broadband via satellite is becoming more common and cheaper, although it still remains a small market.

In 2014, the turnover of the commercial satellite services sector amounted to about 187 billion US dollars, and it has remained strong despite the adverse global economic environment. The fast growth of satellites for communication and broadcasting was favored by the newly competitive choices for video services, coverage in rural areas with diverse service offerings, and extensive sports programming. At present, the demand for satellite services is dominated by North America and Europe. But demand is increasing in emerging markets, which include Latin America, Africa, Central Europe and large parts of Asia. In the next ten years, the growing proliferation of digital content will fuel the demand for satellite infrastructure globally. Some[1] predicts that developing markets like Latin America and parts of Asia will have larger markets for satellite services than North America.

3.1.2 The commercial space infrastructure and support industries

Commercial space infrastructure and support industries include satellite and rocket manufacturing, launch services, ground stations and related equipment. In 2014, total revenues of the sector were 93 billion dollars. It

---

[1] See Northern Sky Research (2017).

represented about thirty three per cent of the turnover of the space commercial sector in 2014.

Satellite manufacturing industry

The global satellite manufacturing sector is composed of satellite and their subsystems manufacturers. The global satellite manufacturing industry growth shows an increasing trend, although with fluctuations from year to year. In 2014, its revenues were about seventeen billion dollars, and represented about seven per cent of total revenues of the commercial space industry.

Chart 3.2 - Satellite Manufacturing Industry Revenues (in billion of dollars)

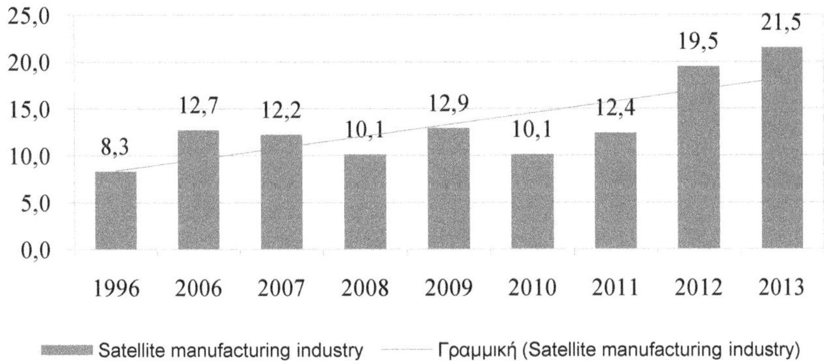

Credit: author. **Source data:** OECD (2014)

Several factors are aiding the expansion of the global satellite manufacturing market. The primary driver was the rising demand and application of commercial communications satellites, due to increasing penetration worldwide.

In 2014, commercial communications satellites manufacturing represented thirty three per cent of the sector's total revenues. Military and navigation satellites were coming next, with military surveillance satellites accounted for thirty eight per cent of the sector's total revenues.

Chart 3.3 - Shares in total revenues

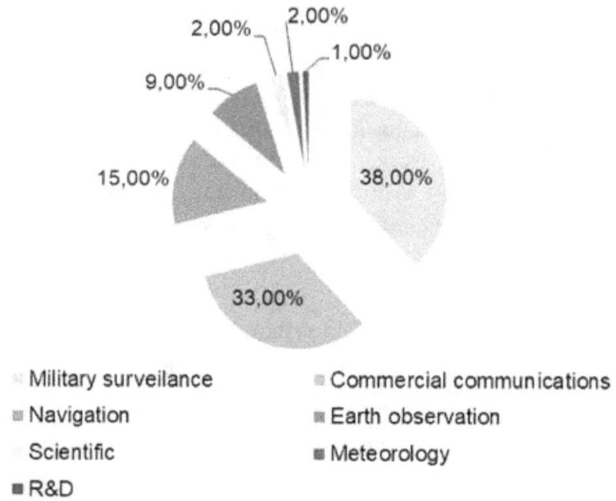

Credit: author. **Source data:** OE CD (2014)

Chart 3.4 - Shares of satellite manufacturing revenues by countries

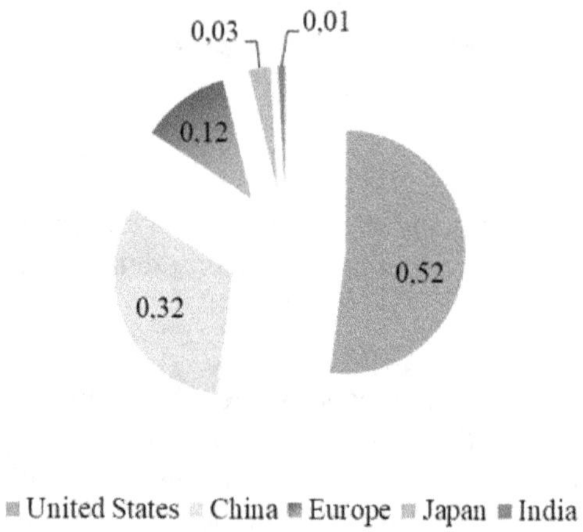

Credit: Author. **Source data:** OECD (2014)

Many economies have developed industrial capacities in satellite manufacturing. The United States has the largest share in the revenues of the satellite manufacturing. It enjoys dominance in the market partly because of strong US government demand, with many institutional satellites. Europe is coming second. The European space industry has seen continual revenue growth. But the European manufacturing industry is dependent on exports for almost half of its revenues, as compared to other industries in Asia and North America. Some of the key market players that are involved in the satellite manufacturing market include Boeing, Indian Space Research Organization, Thales Group, APCO Technologies SA., Lockheed Martin Corp., and Orbital Sciences Corp.

Launching industry

In the early decades of the Space Age, the space agencies of the Soviet Union and United States pioneered launchers' technology, in collaboration with affiliated design bureaus in the Soviet Union and contracts with commercial companies in the United States. All rocket designs were built explicitly for government purposes. After the early 1970s, more countries entered in the launch business. The European Space Agency largely followed the United States model. Other national space agencies, China's CNSA and India's ISRO, also financed the indigenous launchers' construction of their own national designs.

Non-military commercial satellites began to be launched in volume in the 1970s and 1980s, but launch services were supplied exclusively with launch vehicles that had been originally developed for the various national civil and military programs, with attendant higher cost structures. In the following period up to approximately 2010, only eight countries had the technology and facilities to carry out an orbital space launch, or to maintain a fleet of operational launchers: the United States, the Russian Federation, the European Space Agency, China, Japan, India, Israel, Iran and Korea. There were six companies able to commercially launch satellites to geostationary orbit. They include the European Arianespace company, the Russian Federation's International Launch Services, the United States' Lockheed Martin and Boeing, China Great Wall, and Sea Launch, an international consortium (Norway, Russian Federation, Ukraine, and United States). Other companies can launch satellites in lower orbits.

As most institutional satellites were placed into orbit by national launchers, little market competition emerged inside any national market prior to approximately 2010. Some global commercial competition arose between the national providers for international commercial satellite launches. Within the United States, as late as 2006, the high cost structures built in to government contractor's launch vehicles left little commercial opportunity for US launch service providers, but considerable opportunities for low-cost Russian and Chinese boosters based on leftover Cold War military missile technology.

3.1.3 Recent policies changes

One of the main constraints to the growth of the space economy is the high cost of accessing space. In the past, NASA tried to reduce the cost of accessing space with the space shuttle. During early shuttle R & D program there was great debate about the optimal shuttle design to best balance capability, and operating costs. Ultimately a design was chosen, using a reusable winged orbiter and solid rocket boosters, and an expendable external fuel tank. But early expectations of reducing the costs of accessing space did not materialize. The total cost of the actual 30-year service life of the shuttle program through 2011, adjusted for inflation, was 196 billion dollars (in 2010 dollars). Costs per-launch are measured by dividing the total cost over the life of the program by the number of launches. With 135 missions, this gives approximately 1.5 billion per launch over the life of the program. These high costs and several accidents were the main reasons for abandoning the shuttle programs.

Since then, NASA has transitioned from a government owned and operated delivery system to the International Space Station to a privately owned and operated one with multiple competitors, thereby braking the monopoly of Lockheed Martin and Boeing. The target was for the United States to become again independent in human space flights to Earth orbit. NASA also began a multi phase space technology program funded by the United States government and administered by NASA. The program's main goal is to foster research and development into human space flight concepts and technologies in private companies.

NASA has also changed the way of contracting with companies. Back in the early days of the Space Race, people had no idea how rockets would cost. So NASA used what it called "cost-plus" contracts that paid companies the cost of making a rocket, plus either a fixed fee or a percentage. Profit-wise, this was a pretty sweet deal for companies like Lockheed Martin and Boeing, since they did not get to own the rockets or the technology they built, but NASA did. Nowadays, the agency uses "fixed-price" contracts. The companies quote NASA a price, and the space agency does not have to pay them more if they end up going over that number. NASA then signs commercial, fixed price contracts with private companies to deliver cargo to the ISS. With the fixed-price contracts, NASA has less ability to dictate the design or the way the companies are doing things as long as the companies meet the NASA requirements. Now, the companies own the design; they actually own the vehicles. NASA is just renting rides to the International Space Station. And because the companies now get to hold onto their intellectual property, they have the opportunity to commercialize their rockets domestically and internationally, and eventually take civilians on-board.

3.1.4 Launching industry's developments

In the United States, NASA policies stimulated the entrance of new companies in the launching industry. As a result, two new launch vehicles (SpaceX's Falcon 9 and Orbital's Antares), and two new delivery spacecraft (Dragon and Cygnus) for cargo were successfully built. These start-ups have made progress in design and construction of efficient and cheaper launching vehicles. Their cost advantages are not derived from advanced engines. They derive from their systems design and vertically-integrated production, efficient supply chains, and efficiency in operations. This is an example of entrepreneurs reforming or revolutionizing the pattern of production by reorganizing an industry to cope with a problem independently presented by an economic situation (high costs of reaching orbit).

Table 3.1 shows the costs of heavy launchers of the main space faring nations, including a startup, SpaceX.

Table 3.1 – Costs of the Heavy Launchers to geostationary orbit

Vehicles	Ariane 5G	Long March 3B	Proton	Atlas V	Falcon 9	Falcon Heavy
Country	Europe	China	Russia	United States	United States	United States
Load capacity in kg	6,800	5,200	4,630	3,750	8,300	26,700
Launch costs in millions US$	165	60	85	100	62	90
Costs per kg in US$	24,265	11,539	18,359	26,667	7,470	13,280

Credit: author. **Source of data:** FAA Office of Commercial Space Transportation and official SpaceX website www.spacex.com
Note: Data for Falcon 9 refers to 2016, prior to the success in recuperating the first stage of the rocket.

SpaceX, is an American aerospace manufacturer and space transport services company. SpaceX started with the smallest useful orbital rocket[2] (Falcon series), instead of building a more complex and riskier launch vehicle. It was the first privately funded company to successfully launch, orbit, and recover a spacecraft (Dragon[3] in 2010); and to send a spacecraft to the International Space Station (Dragon in 2012).

In the meantime, SpaceX developed methods for reusing the primary stage of the rockets. Previously, the primary stage of the rockets, responsible for the initial thrust through Earth's atmosphere, simply detached after launch and fell into the ocean as waste. Successful experiments have been carried out by SpaceX to recuperate and reuse the primary stage of a rocket. Depending on the amount of work for the maintenance of the primary stage of the rocket after landing, costs per kilogram of payload has been reduced by about 7,500 dollars. After the successful landing of the refurbished Falcon 9 rocket, the company's next goal is for 24-hour turnaround. Moreover, SpaceX is working to build a rocket on which both the first and second stages are reusable, which would further reduce the payload costs.

In 2011, SpaceX began the development of a heavy lifter, Falcon Heavy, configured using a cluster of three Falcon 9 first stage cores, with a total twenty seven Merlin engines and propellant cross-feed. The Merlin engines would produce 22,819 kN (kilonewtons) of thrust at sea level, and 24,681 kN in space. The first stage would be capable of lifting 63,957 kilograms to LEO. When SpaceX finishes its development works, and the rocket is launched, Falcon Heavy will be the world's most powerful rocket in operation. SpaceX is aiming for the first demonstration flight of the Falcon Heavy at the end of 2017.

Competition is producing pressure on other aerospace companies in the United States to review their technologies and reduce costs. Facing direct market competition from SpaceX, the large US launch provider United Launch Alliance (ULA) announced strategic changes in 2014 to restructure its

---

[2]The Falcon rocket series is a family of multi-use rocket launch vehicles developed and operated by SpaceX. The rockets in this family include the flight-tested Falcon 1 and Falcon 9. The Falcon 1 is a small, partially reusable two stage rocket capable of placing several hundred kilograms into low earth orbit. The Falcon 9 v1.2 is an upgraded of Falcon 9 v1.1 with a larger liquid oxygen tank, loaded with subcooled propellants to allow a greater mass of fuel in the same tank volume. The second stage was also extended for greater fuel tank capacity. These upgrades brought a 33% increase to the previous rocket performance.

[3]Dragon is a reusable spacecraft. It consists of a nose-cone cap that jettisons after launch, a conventional blunt-cone ballistic capsule, and an unpressurised cargo-carrier trunk equipped with two solar arrays. The Dragon capsule is equipped with 18 Draco thrusters. It can transport about 3 tons of cargo.

launch business, facing stiff competition from SpaceX also for the launch of US military payloads. As of 2015, ULA faced no competition for the launches of military payloads. In 2015, SpaceX upset the traditional military space launch arrangement in the United States, which has been called a monopoly by many space analysts and criticized by some in the US Congress. As of May 2015, the SpaceX Falcon 9 was certified by the United States Air Force to compete to launch many of the expensive satellites, which are considered essential to US national security.

Europe has remained tied to the old model, characterized by state procurement contracts and a monopolistic launching industry. At the moment, the only company in Europe producing medium-heavy rockets (Series Ariane) is Arianespace. It provides medium-heavy rockets for the launch of commercial satellites. It has a close relationship with the European Space Agency for the launch of ESA satellites and its scientific missions. But competition from the United States, Russia, and China risks marginalizing the old continent. Continuing to face stiff competition on price, seven European satellite operator companies, including the four largest in the world by annual revenue, requested that ESA find immediate ways to reduce Ariane 5 rocket launch costs and, in the longer term, make the next-generation Ariane 6 vehicle more attractive for telecommunications satellites.

3.2 Public administrations engaged in space activities

Public administrations engage in space activities because of the belief that fundamental research would produce new knowledge essential to the health, security, and quality of life of their citizens. Since the start of space efforts, national governments in the United States, the Soviet Union, and Europe have given high priority to the support of science done in and from space. From modest beginnings, space science has expanded under government support to include multibillion-dollar exploratory missions in the solar system and beyond, and space infrastructure.

Public administrations include both civil and defense related organizations. The use of space for defense purpose remained important thanks also to space technologies that have both civil and military applications like weather forecast, which can also be used for early warning, remote sensing, with its applications in intelligence, and global navigation, a precision targeting system. But military budgets are usually spread across several government agencies, which makes them sometimes difficult to track in national accounts. That is the reason why this section concentrates on space agencies.

3.2.1 Space agencies

During the 1950s and 1960s, Europe and Japan were rebuilding their economies after the catastrophes caused by Second World War. Consequently, they were absent from space exploration activities. In the late 1960s, several European countries looked enviously at the American space exploration program and its effects on technological innovation. In 1975, a group of European countries created the European Space Agency (ESA), a body independent of the European Union[4]. The main goal of the ESA is to enable Europe to become and stay competitive in space technology, and to carry out scientific research. ESA coordinates space programs of the participating European countries. The main European launch vehicle Ariane 5 is operated through Arianespace with ESA sharing in the costs of launching and of developing launch vehicles. It maintains a major space-port, the Guiana Space Centre at Kourou, French Guiana.

In the late 1962, Japan also entered in the realm of space exploration. The Japanese government created three agencies: National Space Development Agency (NASDA); Japanese Aerospace Laboratory (NAL); and Institute of Science Astronautics (ISAS). Promoting space research and technological innovation in Japan was the main goals of these institutions. In 2003, the government merged the three agencies into a single one, the Japanese Space Agency (JAXA). The entry of China into space exploration in the early 2000s was an important motivation for the creation of JAXA. JAXA handles research, technology development, and satellite launching. It is involved in many advanced projects. These range from research and development of a space-based solar powered satellite, asteroid exploration, and possible manned exploration of the Moon.

The government of India founded ISRO in 1969 within the Department of Atomic Energy. Promoting space technologies and their applications was the goal of this organization. ISRO developed two rockets: the Polar Satellite Launch Vehicle for launching satellites into polar orbits; and Geosynchronous Satellite Launch Vehicle for placing satellites into geostationary orbits. These rockets have already launched several communications and earth-

---

[4]ESA is the first example of international cooperation on a regional level. Its activities are grouped into two categories: mandatory and optional programs. The programs carried out under the general budget and scientific programs are mandatory. These programs are funded by member states in proportion to the size of their gross domestic products. Member countries are free to decide on their participation in optional programs. Optional programs include projects such as Earth observation, telecommunications, space transportation, and manned space flight. For example, the International Space Station and research in micro gravity are funded under optional programs.

observation satellites. In 2008, ISRO sent a mission to the Moon, Chandrayaan-1. In 2013, it launched the Mars Orbiter Mission (MOM). MOM is a "technology demonstrator" project to develop technologies for design, planning, managing, and operating interplanetary missions. It carries five instruments to advance knowledge of Mars, a secondary scientific goal. As MOM successfully entered Mars orbit on 24 September 2014, India became the first Asian country to do so.

China's space program began in 1956 as part of Sino-Soviet cooperation. China began a rudimentary ballistic missile program in response to perceived American threats. After the 1960 Sino-Soviet crisis, China continued its space program independently from the Soviet Union. The Central Military Commission controlled this program to meet national defense needs. In 1990, the Chinese government reorganized its space program along with other defense-related industries. Initially it established two agencies: China National Space administration (CNSA) and China Aerospace Corporation (CASC). The former was responsible for policy, the latter for execution. This arrangement proved somewhat unsatisfactory as these two agencies were, in reality, one large agency sharing both personnel and management. After a new reorganization, the space program is now the responsibility of CNSA and the Science, Technology and Industry Commission. The government split CASC into smaller state-owned companies. CNSA contracts out their operational requirements to entities that are government-owned but not government-managed.

Over the years, other nations initiated space activities, although with smaller programs. As of 2017, seventy one different government space agencies are in existence. Thirteen of those have launch capability. Six government space agencies - the Indian Space Research Organization (ISRO), the European Space Agency (ESA), the China National Space Administration (CNSA), the Japan Aerospace Exploration Agency (JAXA), the National Aeronautics and Space Administration (NASA) and the Russian Federal Space Agency (RFSA or Roscosmos) have full launch capabilities. These include the ability to launch and recover multiple satellites, deploy cryogenic rocket engines, and operate extraterrestrial probes. Only three currently operating space agencies in the world - RFSA and the CNSA and NASA - are capable of human space flights as of 2017.

In 2017, the total budget of the main space-faring nations amounted to about 42 billion dollars (Chart 3.5). The space agencies' budgets of the United States, Russia, ESA, Japan, India, and China accounted for about 76 per cent of the total.

Chart 3.5 - Space budgets of main space faring nations
(in billion of dollars)

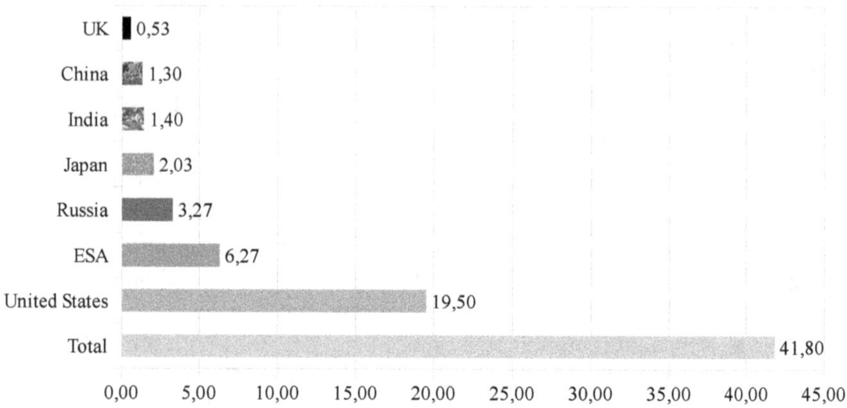

Credit: author. **Source of data:** National space agencies statistics

While total dollars count, how much of a country's economy is invested in space activities is also relevant, showing the relative priority of space in overall governmental policies.

Chart 3.6 - Space budget of main spacefaring nations
(in per cent of GDP)

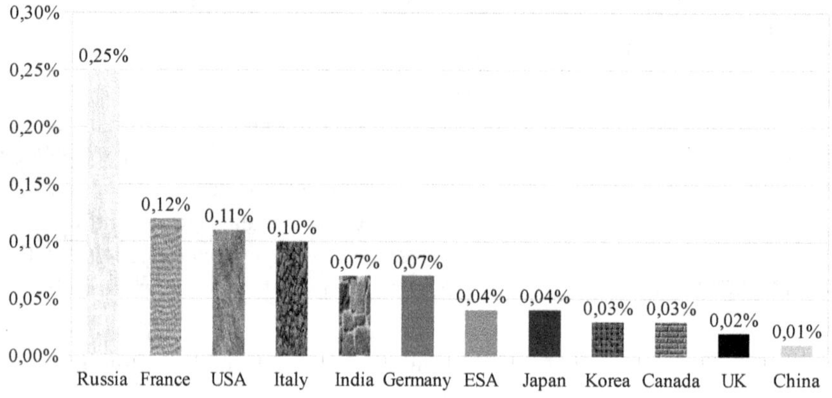

Credit: author. **Source of data:** National space agencies and IMF statistics.

By share of GDP, Russia takes the lead. It is followed by France, the United States, and Italy. India moves up the rankings before Germany, Japan, ESA, Korea, and China[5]. India takes space activities seriously as a part of its economy.

Finally, attention should be given to the shares of human space flight (HSF) within the total budgets of the national space agencies. NASA and RKA (the Russian space agency) are unique in spending high amounts of their budgets on HSF, in the range between 35-45%. Meaningful figures for historical or projected spending on HSF by China are virtually impossible to get. But Chinese space-station programs and manned-space flights show the high priority given by the Chinese to HSF. The Japanese, Canadian and European shares for HSF in their total spending lie between ten and twenty per cent and have even been declining. While India has ambitious plans, its HSF spending is still a tiny fraction (1%) of ISRO's overall budget. The investment portfolio priorities of space agencies, except for NASA, RKA, and the Chinese Space Agency, are centered on science projects, technology, and satellite activities.

3.2.2 Activities of space agencies

Space agencies' main activities have concentrated on robotic solar system exploration, universe exploration through space telescopes, R&D programs, and building of space infrastructure. The main goals were: a better understanding of the solar system, including the search for life forms; understanding the many objects and phenomena in the universe that are better observed from a space perspective; and using for human benefit the resources and attributes of the space environment. Human space exploration has been limited to suborbital flights and the short term permanence of astronauts at the International Space Station.

Solar system exploration

By 2017, all the planets of the solar system have been visited by robotic spaceships[6]. Mercury remains the least explored of the inner planets. Venus was the first target of interplanetary flyby and lander missions and, despite being one of the most hostile surface environments in the Solar System, has had more lander sent to it than any other planet in the Solar System. Its thick atmosphere traps heat in a runaway greenhouse effect. Radar from space-

[5]Meaningful figures for historical spending on the Chinese space program are virtually impossible to obtain. The estimates of civilian-space spending are derived from Western expert reports, showing that CNSA spending is around one tenth of NASA spending.
[6]A timeline of solar system exploration can be found in:
https://en.wikipedia.org/wiki/Timeline_of_Solar_System_exploration.

crafts showed that Venus has evidence of volcanism, and impact craters, but no plate tectonics, which may contribute to its runaway greenhouse effect.

Mars have been visited by many fly-by, orbiting, and lander missions[7]. Probes sent from Earth, beginning in the late 20th century, have yielded a dramatic increase in knowledge about the Martian system, focused primarily on understanding its geology, past evolution, and habitability potential. There is mounting evidence from orbiting and lander missions that in the past Mars had a watery environment. But how Mars may have transitioned from a more watery environment to the dry and dusty climate it has today, and whether life could have emerged in that environment is still unknown. To find out, one needs to understand how geologic, climatic, and other processes have worked to shape Mars and its environment over time, as well as how they interact today. Future missions may answer these questions, which may be important for future human exploration of Mars.

Several asteroids and comets have now been visited by probes. The asteroid's missions studied the asteroids' shape, spin, topography, color, composition, density, and history. Asteroid missions are important for our understanding of how the solar system evolved. But the analysis of their physical characteristics is important for future human visits to these space rocks to extract water and metals. Comets might have originally brought some of the water and organic molecules to Earth that now make up life here. To research this hypothesis, the Rosetta mission, which landed a probe on a comet on Nov. 12, 2014, is studying its nucleus and environment, observing how it changes as it approaches the sun.

Probes and fly-by allowed to study the moons[8] and the rings of the outer planets and discovered previously unknown moons orbiting them. In the case of Jupiter, the Galileo orbiter contained a small atmospheric probe, which separated from Galileo and entered the Jovian atmosphere. After using a heat shield to brake, it popped open a parachute and descended through the clouds of Jupiter's upper atmosphere for fifty eight minutes, allowing an analysis of Jupiter atmosphere. Saturn has been explored only through unmanned spacecraft launched by NASA. These missions consist of flybys in 1979 by

---

[7]The exploration of Mars has been an important part of the space exploration programs of the Soviet Union (later Russia), the United States, Europe, Japan and India. Dozens of robotic spacecraft, including orbiters, lander, and rovers, have been launched toward Mars since the 1960s.

[8]They discovered the volcanic activity of Io, one of Jupiter's moon, revealed that Io's surface is constantly being remade by volcanic forces. They provided strong evidence that underneath Europa's cracked icy surface is a vast saltwater ocean, an important prerequisite for life.

Pioneer 11, in 1980 by Voyager 1, in 1982 by Voyager 2. After having visited Saturn successfully, it was decided to continue and fund further missions by Voyager 2 to Uranus and Neptune. A probe (Cassini mission) was sent in 1997 to study the planet Saturn and its system, including its rings and natural satellites. In 2017, at the end of its mission, the Cassini spacecraft executed the "Grand Finale" of its mission: a number of risky passes through the gaps between Saturn and Saturn's inner rings, before the spacecraft was destroyed.

Space telescopes

Most of the information that man is able to obtain about the universe comes to him in the form of electromagnetic radiations. Stars concentrate most of their energy in a relatively narrow band stretching from the near ultraviolet into the infrared portion of the spectrum. One can see into these hidden parts of the spectrum by using various kinds of detectors. But, while scientists have developed sensors that will enable us to see and measure much of this very short and very long wavelength energy, astronomers have not been able to exploit them fully, because only part of the energy in these portions of the spectrum reaches the Earth. Much of it is blocked or deflected, absorbed or screened out by the Earth's atmosphere and magnetic field.

Space telescopes allowed to overcome those difficulties. They are in essence large satellites, which allow astronomers to answer many questions. A prime question concerns the composition of the universe, i.e the relative abundance of different atoms[9]. Second, a unique window on the Universe enabled by space telescopes allows creating images of small patches of sky that are the deepest ever obtained at optical wavelengths. The images reveal galaxies billions of light years away, and have generated a wealth of scientific papers, providing a new window on the early Universe. Space telescopes allow also to measure distances to Cepheid variable stars more accurately than ever before, and thus constrain the value of the Hubble constant, the measure of the rate at which the universe is expanding, which is also related to its age[10]. Third, space telescopes are used in the discovery of extrasolar planets in the Milky Way. Photographing these objects from space, and the possibility of finding

---

[9]The most abundant elements in the universe are thought to be the lighter ones. Unfortunately, their strongest spectral lines are in the far ultra violet region of the radiation spectrum, which can never be observed from the surface of the earth, but they can be observe by space telescopes.

[10]While helping to refine estimates of the age of the universe, they also cast doubt on theories about its future. Astronomers from the High-z Supernova Search Team and the Supernova Cosmology Project used ground-based telescopes and HST to observe distant supernovae and uncovered evidence that, far from decelerating under the influence of gravity, the expansion of the universe may in fact be accelerating.

new ones, is an important result. It would permit to make detailed observations of the structure of the surfaces of these planets and the composition of their atmosphere.

Space agencies R&D programs

Despite meager resources, space agencies have been financing research and development programs on new propulsion schemes. The goals are: shortening astronauts' time in space; and reducing propellant mass. When discussing the efficiency of propulsion, engineers focus on how effectively it uses reaction mass11. Specific impulse, i.e., thrust deliverable by ejecting a unit weight of propellant, is one way of measuring propulsion efficiency. It gives us a quick way to calculate the rocket thrust, if the weight flow rate through the nozzles is known. The higher the thrust, the higher is the spaceship's change in velocity. Second, it measures the engine efficiency. If two different rocket engines have different specific-impulse values, the engine with higher specific impulse is more efficient because it produces more thrust for the same propellant mass.

Table 3.2 shows specific impulses for different spaceship's engines.

Table 3.2 – Specific Impulses in Seconds

Type of spaceship engine	*Specific impulse*
Solid fuel	250
Liquid fuel	450
Ion	3,000
Nuclear fission	800-10,000
VASIMIR plasma	1,000-30,000
Nuclear fusion	2,500-200,000
Nuclear pulse	10,000-1 million
Matter-Antimatter	1 million-10 millions

Credit: author

Chemical propulsion has the lowest specific impulse. Electrical propulsion uses electrical energy to change the spaceship velocity. It works by electrically expelling the propellant at high speed. The electric propulsion has a specific impulse of around three thousand seconds. Due to limited electric power, the

[11]Reaction mass is the mass against which spaceships operate to produce acceleration. In a rocket, for example, the reaction mass is the fuel shot backwards to create propulsion.

thrust is much weaker compared to chemical rockets, but since electric propulsion can offer a small thrust for a long time, it can achieve high speed. So, they are optimal for robotic exploration of the solar system. Electric propulsion, already tested successfully on many robotic mission, is a mature technology.

Nuclear propulsion produces high effective exhaust velocity, so it reaches higher specific impulses. Nuclear propulsion can overcome chemical rockets' specific impulse limitations because the energy source and the propellant are independent of each other. Compared with chemical and electrical rockets, nuclear propulsion is more reliable and flexible for long-distance missions. It can achieve space missions at a lower cost. In a nutshell, the reason for these advantages is that it can get "more miles per gallon" than chemical and electrical rockets. Research on nuclear fission propulsion is well advanced in the United States[12], Russia[13], and Europe.

More advanced research on nuclear-fission propulsion is being conducted in the United States and Europe. We shall mention two nuclear-fission propulsion engines here. The first engine, developed in the United States, is called the fission-fragment engine[14]. The fission-fragment engine directly harnesses hot nuclear-fission products for thrust as opposed to using a separate fluid as a working mass. Prof. Rubbia and his colleagues advanced a

---

[12]In the United States, research on nuclear fission propulsion began in 1961 with the NERVA program (Nuclear Engine for Rocket Vehicle Applications). During testing in the laboratory, the NERVA nuclear engine reached a specific impulse of eight hundred and fifty seconds. However, due to the general lack of interest in human space exploration at the end of Apollo project, NASA canceled the NERVA program in 1972. After 1972, research on nuclear fission propulsion continued in the United States and Europe. Research and technology efforts at NASA focused on nuclear thermal rockets and on electric nuclear propulsion. In a nuclear thermal rocket, a working fluid is heated to a high temperature in a nuclear reactor. The heated fuel then expands through a rocket nozzle to create thrust. In a nuclear electric rocket, nuclear thermal energy is changed into electrical energy that is used to power one of the electrical propulsion technologies.

[13]The Soviet Union had its own nuclear propulsion program. The Russians tested the nuclear rocket RD-0410 near the nuclear test site at Semipalatinsk. They foresaw its use in the Kurchatov 1994 manned Mars mission. budget constraints and the lack of interest by the Soviet authorities for interplanetary missions put a stop to nuclear propulsion research.

[14]This engine is called a fission-fragment engine. The Idaho National Engineering Laboratory and Lawrence Livermore National Laboratory developed it. The fuel is placed into several very thin carbon bundles, each one sub-critical. Bundles are collected and arranged as spokes on a wheel. Several wheels are stacked on a common shaft to produce a single large cylinder. The entire cylinder is rotated so that few bundles are always in a reactor core. Surrounding fuel makes the bundles go critical.

second proposal, incorporated in Project 242[15]. This propulsion engine works on the principle of direct transfer of energy from the fission fragments to a gas to generate thrust. These two most advanced engines can reach a specific impulse of 30,000 seconds and more. These more advanced propulsion engines are optimal for human exploration of the near-Earth planets, such as Mars, because they reduce travel times and propellant mass. Although these engines are at the conceptual level, they are starting points for building nuclear-powered spaceships for human Mars exploration.

Rockets that harness the power of nuclear fusion offer the next big leap in the quest to explore vast swathes of the solar system. A nuclear fusion rocket has the highest specific impulse with the exception of matter-antimatter engines. It would provide efficient and long-term acceleration in space without the need to carry a large fuel supply. Nuclear-fusion rockets would slash travel times. NASA has pursued research into the nuclear fusion propulsion. It has funded several early-stage fusion ideas via a program called NIAC (NASA Institute for Advanced Concepts). NASA's Glenn Research Center has proposed a small ratio spherical torus reactor, for its Discovery II conceptual vehicle design[16]. The main alternative to magnetic confinement is inertial-confinement fusion[17] (ICF). A small pellet of fusion fuel is ignited by an electron beam or a laser. To produce direct thrust, a magnetic field will form the pusher plate. A new approach, the magnetic-target fusion[18] (MTF), combines the best features of the magnetic confinement and inertial confinement fusion approaches.

Fusion rockets will not be powering a spacecraft soon. Fusion reactors' large mass is the main disadvantage for fusion powered spaceships. Recently, Lockheed Martin's research unit announced intentions to build and test a compact fusion reactor in less than a year[19]. One of the project's innovations is using superconducting magnets that produce strong magnetic fields with less energy than conventional magnets. According to the research director, the reactor's architecture reduces its mass by a factor of ten with respect to

---

[15]See Bignami G. et al. (2011).

[16]See Craig H. W. (2005).

[17]In the 1980s, Lawrence Livermore National Laboratory and NASA studied an ICF powered "Vehicle for Interplanetary Transport Applications" (VISTA). The conical VISTA spaceship could deliver a 100 tonne payload to Mars orbit and return to Earth in 130 days. VISTA needs forty one tonnes of deuterium/tritium plus 4,124 tonnes of hydrogen for this trip. The exhaust velocity was estimated at 157 km/s. The estimated specific impulse is around 70,000.

[18]The NASA Human Outer Planets Exploration (HOPE) group has investigated a manned MTF propulsion spacecraft.

[19]See http://www.lockheedmartin.com/us/products/compact-fusion.html.

classical fusion reactors. If research were to be successful and costs contained, it will pave the way not only to produce electricity but to be practical on spaceships. The reactor's design is at an early stage. But one piece of good news already exists: if the cliché that nuclear fusion research will make s commercial reactors available in thirty years, this means that we have now to wait only five to ten years.

Earth observation

Most space agencies have launched Earth observation satellites for environmental monitoring and meteorology. Major programs of Earth observation are carried out by NASA, ESA, national European space agencies, the Russian space agency, Japan space agency, and China space agency. But a number of other countries are also carrying out Earth observation programs. NASA, frequently in collaboration with the National Oceanic and Atmospheric Administration, has up to now the most extensive Earth observation program. NASA's Earth Observing System (EOS) is a coordinated series of polar-orbiting and low inclination satellites for long-term global observations of the land surface, biosphere, solid Earth, atmosphere, and oceans. The EOS Project Science Office is committed to bringing program information and resources to the Earth science research community and the general public alike. NASA oversees the procurement, design, and manufacture of Geostationary Operational Environmental Satellite system (GOES). GOES is operated by the United States' National Oceanic and Atmospheric Administration to supports weather forecasting, severe storm tracking, and meteorology research.

ESA's Earth Explorer missions are designed to improve our understanding of Earth. They use cutting-edge space technologies to learn more about the interactions between the atmosphere, biosphere, hydrosphere, cryosphere and Earth's interior, essential how Earth works as a system. The Sentinel missions are developed specifically for the European Union's Copernicus program. It is the largest environmental monitoring program in the world. ESA has also been instrumental in setting up a number of initiatives such as the Climate Change Initiative and the International Charter 'Space and Major Disasters'. The Climate Change Initiative makes full use of Europe's Earth observation space assets to generate long-term data sets on key indicators of climate change. These 'essential climate variables' provide Europe with a powerful tool to monitor the climate system and help predict the effects of change. The International Charter provides rapid access to satellite data to help disaster management authorities in the event of a disaster such as a flood, earthquake or volcanic eruption.

Environmental satellites can assist environmental monitoring by detecting changes in the Earth's vegetation, atmospheric trace gas content, sea state, ocean color, and ice fields. By monitoring vegetation changes over time, droughts can be monitored by comparing the current vegetation state to its long term average. Environmental satellites carry instruments that are operated at a relatively low altitude. To get (nearly) global coverage with a low orbit it must be a polar orbit or nearly so. A polar orbit is one in which a satellite passes above or nearly above both poles. Near-polar orbiting satellites commonly choose a Sun-synchronous orbit, meaning that each successive orbital pass occurs at the same local time of day. This can be particularly important for applications such as remote sensing atmospheric temperature, where the most important thing to see may well be changes over time which are not aliased onto changes in local time.

A weather satellite is a type of satellite that is primarily used to monitor the weather and climate of the Earth. But these satellites see more than clouds and cloud systems. City lights, fires, effects of pollution, auroras, sand and dust storms, snow cover, ice mapping, and boundaries of ocean currents are other types of environmental information collected using weather satellites. Meteorological satellites mainly use a geostationary orbit. A geostationary orbit allows uninterrupted coverage of more than 1/3 of the Earth. Three geostationary satellites at longitudes separated with 120 degrees can cover the whole Earth except the extreme polar regions.

3.3 International collaboration

During the Cold War, major scientific and engineering breakthroughs took place in different parts of the world, often in isolation, as military research and development, and industrial secrecy forced the super powers to preserve their own technological advances. After the fall of the Berlin Wall and the end of the Cold War, there was an increase in collaboration and dissemination of scientific advances, knowledge flows and dual-use technological transfers from OECD countries and the Russian Federation to other parts of the world. Joint institutional space programs provided an excellent way to develop and use national expertise and scientific capabilities, while sharing financial burdens in common large-scale projects that would have been impossible to launch by individual countries.

3.3.1 International space institutions

All space institutions pay a lip service to international collaboration. In fact, most regional space agreements are directed to promote the nation-state interests. Regional space agreements exist in Asia. Japan lead the Asia-Pacific Regional Space Agency Forum (APRSAF). APRSAF is a voluntary organization consisting of governments, space agencies, the private sector and academia of Asian countries. Its goals are the promotion and expansion of space activities and applications as part of an over whole strategy for the development of the region. A second Asian space agreement is the Asia-Pacific Space Cooperation Organization (APSCO). APSCO is an intergovernmental organization with strong binding political commitments, led by China. Although both organizations recognize that space has dimension beyond national considerations, which can only be addressed along with international partners, they serve both nation-states interests in promoting their technology in the region.

Genuine international space institutions exist, but they are few. One of these is the Committee on Space Research (COSPAR). It was established by the International Council for Science in 1958. Among COSPAR's objectives are the promotion of scientific research in space on an international level, with emphasis on the free exchange of results, information, and opinions, and providing a forum, open to all scientists, for the discussion of problems that may affect space research. COSPAR has created a number of research programs on different topics, a few in cooperation with other scientific Unions. One of these programs is Planetary protection. Planetary protection is a guiding principle in the design of an interplanetary mission, aiming to prevent biological contamination of both the target celestial body and the Earth in the case of sample-return missions. COSPAR meets every two years, in a gathering of 2000 to 3000 scientists. One of its tasks is to develop recommendations for avoiding interplanetary contamination. Its legal basis is Article IX of the Outer Space Treaty.

A second one is the International Space University (ISU). ISU was founded as an international institution of higher learning, dedicated to the development of outer space for peaceful purposes through international and multidisciplinary education and research programs. ISU was founded on the "3-Is" philosophy providing an Interdisciplinary, Intercultural, and International environment for educating and training space professionals and post-graduate students. The ISU faculty members include astronauts, space agency leaders, space engineers, space scientists, managers, and experts in space law and policy comprising an international collection of experts in technical and non-technical space-related fields.

A third one is the International Institute of Space Law (IISL). IISL is an independent non-governmental organization dedicated to fostering the improvement of space law. The objective of the IISL is to promote the expansion of the rule of law in the exploration and uses of outer space for peaceful purposes. To this end, it holds meetings, colloquium and competitions on juridical and social science aspects of space activities, prepares studies and reports, publishes books, proceedings, reports and position papers, and cooperates with appropriate international organizations and national institutions in the field of space law. The IISL is an official observer of the United Nations Committee on the Peaceful Uses of Outer Space (UNCOPUOS) and its Scientific & Technical and Legal Subcommittees.

A fourth one is the International Space Science Institute (ISSI). ISSI is an institute of advanced study where scientists from all over the world meet in a multi- and interdisciplinary setting to reach out for new scientific horizons. The main function is to contribute to the achievement of a deeper understanding of the results from different space missions, ground based observations and laboratory experiments, and adding value to those results through multidisciplinary research in the framework of International Teams, Workshops, Working Groups, Forums or as individual visiting scientists. The program of ISSI covers a wide spectrum of disciplines ranging from the physics of the solar system and planetary sciences to astrophysics and cosmology, and from Earth sciences to astrobiology.

3.3.2 Joint institutional space programs

Joint institutional space programs are bilateral or multilateral programs among space agencies. Their main motivations are the use of different technologies developed by the partners and the sharing of the financial burden of large space programs. The International Space Station (ISS) is a case in point. ISS is a project jointly owned by NASA, Russian RKA, European ESA, Japanese JAXA, and the Canadian space agency. Ownership and use of the space station are established by intergovernmental treaties and agreements.

The history of the international space station is a long one. After the end of the Apollo program, plans for building a space station in low Earth orbit vanished along with Von Braun. Eventually, proposals for a large space station re-emerged in the United States. NASA realized the insufficiency of its own resources, and the proposals remained an expensive paper project. Meanwhile, the Soviets built their own no-frills space station, consistent with their pragmatic space program. The Mir station (Mir in Russian means both "peace" and "world") was the first modular space station. It was assembled in orbit from 1986 to 1996. The station served as a microgravity research laboratory. Crews conducted experiments in plant biology, human biology, and physics and observations in astronomy and meteorology. Mir was a success, but, in addition to its Russian-language operations, it was in an orbit inclined 51° to the equator.

After the Soviet Union's demise, Russia adopted a relatively cooperative foreign policy with the West. Despite economic turmoil, Russia retained advanced technical knowledge of, and capabilities in, space operations. Since NASA was happy to work with Russia's space agency, politicians on both sides jumped on that wonderful opportunity. In 1993, Al Gore and V. Chernomyrdin signed an agreement to design, develop, and use a permanently inhabited civil s station for peaceful purposes. The inevitable price for the United States to pay to work with Russia was to build ISS in an orbit inclined 51° on the equator. To transfer from one orbit to another, a spaceship has to spend energy, i.e., had to burn more fuel. Celestial mechanics stipulates that it is easier to transfer from an orbit inclined by 28° to one inclined by 51° than the reverse. A spaceship can move from orbit "Cape Canaveral" to orbit "Baikonur", but the reverse process is inefficient. This meant that the shuttle would require more energy, but NASA thought they could afford it.

Plans, as always optimistic, foresaw the completion of the ISS by 2003. But it took an extra eight years to complete. Russia launched module "Zaria" ("dawn" in Russian), from Baikonur in 1998: indeed the first launches were all Russian. They used the cargo vehicle "Progress" (36 flights to the ISS). Later on, launches were split halfway between the US Space Shuttle and the Russian spacecraft Soyuz. Recently, Europe and Japan used their own cargo vehicles. Although, in the ISS global economy, cargo business is not a very important element, it has helped to justify the ESA to operate its launch base in Kourou (French Guiana). For the first time, a European vehicle reached the ISS with authorization to dock, having started from a spaceport other than Baikonur or Cape Canaveral.

Since ISS's overall costs amount to around 150 billion dollars[20], tough questions have been raised about the ISS's usefulness. It is an inefficient base for launching spaceships to explore the solar system because of high inclination on the ecliptic plane[21]. But does it have a scientific usefulness? Modest, and not enough to justify the large investments as politicians tried to make the world believe since the Reagan presidency. This has resulted in the justifiable critical reaction of the global scientific community. In other realms, ISS is a

---

[20]This includes: NASA's budget of 58.7 billion dollars; Russia's 12 billion dollars; Europe's 5 billion dollars; Japan's 5 billion dollars; Canada's 2 billion dollars; and the cost of 36 shuttle flights to build the station, estimated at 1.4 billion dollars each, or 50 billion dollars total. Assuming 20,000 person-days of use from 2000 to 2015 by two to six-person crews, each person-day costs 7.5 million dollars totaling around a half billion dollars.

[21]If the sun's path is observed from the Earth's reference frame, it moves around Earth in a path tilted 23.5°. This path is called the ecliptic. Observations show that other planets, except for Pluto, orbit the sun in the same plane. The ecliptic plane thus comprises most of the objects orbiting the sun. A spaceship leaving ISS for interplanetary missions will have to move from an orbit inclined by 51° to one inclined by 23.5° and so require higher fuel consumption. Hence, it is clear the inefficiency of ISS as a staging point for interplanetary missions.

useful laboratory to test 3D printing[22] in weightlessness, and it is thought that printing can be important in future space operations. It is relevant to building in-space infrastructure, plus equipment and other products using raw materials from space. Over the long run, the ability to deliver components on demand without the need of launch vehicles can redefine how space-mission strategies work.

ISS has proven useful as a training ground for further human space flight. So far, several dozen astronauts from many countries have visited it. It enables on-going studies on the effects on the human body of extended periods in space. But these experiments take place in low Earth orbit, i.e., above the atmosphere but within the Earth's magnetic field. The magnetic field blocks the cosmic radiation that otherwise would be hitting the vehicles and the astronauts within. The most serious problems for human interplanetary flight coming from exposure to ionizing cosmic radiation and how to protect from them could not be analysed and revolved.

There is no question that the ISS has taught the United States, Russia, Europe and Japan to develop new technologies and has stimulated industrial and global collaboration. This is ISS' most important role. Construction, assembly, and operation of the International Space Station requires the support of facilities on Earth managed by the partner agencies and industries involved in the program. The International Space Station program brings together international flight crews, multiple launch vehicles, globally distributed launch, operations, training, engineering, and facilities; communications networks; and the international scientific community.

3.4 Impacts on the world economy

Socio-economic impacts derived from space activities are diverse. They range from medium term effects of cost/efficiency gains on productivity and growth to short term economic effects of space agencies' spending on economic growth and employment.

3.4.1 Cost/efficiency gains

Table 3.3 reports some of the cost-efficiency gains of using satellite applications. The table gives a qualitative assessment whether there was likely a cost/efficiency gain in one of the identified sectors.

---

[22]Recently, SpaceX's CSX-4 vehicle brought to the space station a 3D printer, built by the company Made in Space. It started the first 3D printing in space. It took over an hour to finish the job. More experiments will follow in the following months.

Table 3.3 – Selected economic effects of satellite applications

	Weather	Earth observation	Navigation	Communications	
Efficiency gains					
Airline industry	yes	n.a.	yes	yes	Efficiency gains in the air transport sector due to better weather forecasts
Ground transportation	n.a	n.a	yes	n.a	Efficiency gains in moving people and goods.
Sea shipping	yes	n.a.	yes	yes	Efficiency gains, as a result of better weather forecasts, and navigation (GPS, satellite-based ice charts)
Fishing industry	yes	n.a.	yes	yes	Efficiency gains and improved control of resources (illegal fishing)
Agriculture	yes	yes.	yes	yes	Saving water and higher yields
Energy sector	yes	n.a.	n.a	yes	Annual gains as a result of better forecasting demand for electricity
Costs gains					
Oil industry	yes	n.a	n.a.	yes	Cost avoidance/savings in terms of detecting and managing oil incidents
Flood	yes	yes.	yes	yes	Cost avoidance/savings in terms of anticipating and managing flood events
Mining industry	n.a.	yes	n.a.	yes	Savings in exploration.

Credit: author

Although costs/efficiency gains of using satellite applications are often qualitative, in some case monetarily quantifiable in documented cases are possible. Efficiency gains are particularly relevant in the aviation and ground transportation. The economic impacts of the GNSS application on the aviation industry are significant. The Global Market Forecast[23] for the period 2017-2036, done by Airbus, indicates that the air traffic between advanced and emerging air transport markets is expected to increase at an average annual rate of 5.1 per cent. The forecast also indicates that the number of passenger and freight aircraft fleets is expected to have increased by 109 and 82 per cent respectively. The improvement if airspace management procedures would help make use of the available infrastructure to the full, by contributing to safety and accessibility. GNSS relieves bottlenecks and improve safety and efficiency to deliver more on-time and fuel-efficient flights. Consumers benefit from better travel experiences through the reduction in delays resulting from on-time performance by controllers and operators. The FAA expects that GNSS improvements will reduce overall flight delays in the U.S. by 41 per cent between 2013 and 2020.

The ground transportation systems of cars, buses, metros, and rails are required to be efficient and safe to move workers and goods around the city smoothly. Construction of the intelligent technology systems (ITS) provides cities with cost-effective solutions to maximize the economic returns of the city infrastructure to build safe, smart, and efficient transportation systems. The U.S. Government Accountability Office (GAO) conducted a literature review[24] to identify cost-and-benefit studies of real-time traffic and information systems and technologies. The GAO report concludes that the benefits are greater than the costs. The study found that the present value of total cost savings is about 30.2 billion dollars (generating from the benefits to mobility, the environment, and safety) compared to the present value of the costs of 1.2 billion dollars of the program.

Efficiency gains are also present in agriculture. Decades of satellite observations have enabled scientists to address fundamental questions about weather and climate. Satellite observations yield continually updated knowledge of the state of the atmosphere, helping meteorologists to devise models that project the weather with much improved accuracy compared to previous forecasts. A recent study shows that the meteorological-satellite services help to reduce water use and to obtain higher yields per hectare of cultivated land. The economic impact is estimated at forty two hundred and ten dollars per hectare. Heavy equipment can use GNSS in precision agriculture. This is use-

---

[23]See Airbus (2016).

[24]U.S. Government Accountability Office (2012)

ful for row crop operations and when spraying. Harvesters with yield monitors can also use GNSS to create a yield map of the paddock being harvested.

Satellite communication services (via the Internet and mobile phones) creates efficiency and economic opportunities by reaching areas of the world that are difficult otherwise to reach. Access to information and communication infrastructure greatly enhances economic growth. When a reliable and affordable medium for information exchange is available, previously unanticipated developments ensue. Most areas in developing countries are sparsely populated and highly rural. Satellite communication is an excellent option for meeting this and many other pressing communication needs of developing economies. Developing economies are actively implementing data networks, and have undertaken policy reforms to facilitate such networks. This is due to the undeniable fact that optimizing satellite communications can accelerate national development.

Cost gains can be achieved by satellite applications. Satellite earth observations work through processes such as the search for new sources of raw materials in inaccessible places, and in monitoring of extreme events. Search for new sources of raw materials in inaccessible places helps in reducing costs of exploration. Timely information on extreme events helps in reducing costs in terms of human life. Also, as a result of satellite-based observations, pollution is now viewed as a global, not a local, phenomenon. Information from satellites provides crucial data to inform models of pollution dynamics and helps scientists predict changes in the atmospheric composition with greater confidence. For example, current understanding of climate change has been improved by basic satellite-based research on sea surface temperature, ice sheet flows, the El Niño-Southern Oscillation, and Earth's carbon cycle and energy fluctuations. Finally, satellite applications can result in cost avoidance/savings to the oil industry by detecting and managing oil incidents.

A note of warning is necessary. It is important that findings, data cited, and conclusions are placed in context. All previous and cited studies on this subject were conducted using limited and selected data samples, and that the findings from these samples cannot be extrapolated to encompass, explain, or make predictions concerning the full range and scope of benefits. The estimates are provided by a variety of sources, each using different methods to gather and report the information. Development of common methodologies is necessary to improve the above analysis. At present, works in this direction are done at the OECD, which can improve further the analysis.

3.4.2 Economic impact of space exploration

Space exploration missions have direct economic effects via the multiplier effect, and spin off of technological innovations of space programs on longer term economic growth. Chase Econometric Associates conducted an econometric investigation[25] of the relationship between NASA expenditures and the U.S. economy. The first phase of the Chase study employed the 185 inter industry input-output model developed at the University of Maryland to analyses the short-run economic impact of NASA R&D expenditures. Simulations of the input-output model were undertaken assuming that one billion dollars of federal expenditure was transferred (proportionately) from other nondefense programs to NASA with no change in the size of the federal budget. Chase estimated that the one billion dollars transfer would increase manufacturing output by one percent, and would increase manufacturing employment by 20,000 workers.

But the hypothesis behind this study was no changes in the federal budget. Allowing deficit spending may change the above results. There are several channels through which government policies can affect economic growth. In the economic literature, there are three competing hypothesis of this effect. The first one is the Pre-Ricardian equivalence proposition, which states that deficit spending has no effect on economic growth. The second one is the Keynesian multiplier, which states that deficit spending has a positive effect on economic growth and private investments. The third hypothesis states that deficit spending crowds out private investment and reduces economic growth. Empirical tests of these hypotheses seem to confirm the third one. But the degree of crowding out is not independent from the way the deficit is financed. The higher the amount of bond financing, the higher is the effect on real interest rates and hence the greater the scope of crowding out of private investments. Allowing deficit spending would reduce the positive effects of NASA spending on the economy.

Technological spin off from space exploration are many. PS300, a solid lubricant coating was developed by NASA. It is deposited by thermal spraying which protects foil air bearings. PS300, with its ability to reduce friction and emissions, has been widely used in industrial applications such as refrigeration compressors, turbochargers, and hybrid electrical turbo generators. NASA, along with Goodyear and Rubber company, developed a fibrous material - five times more durable than steel. It was used in parachute shrouds to soft-land the Viking Lander spacecraft on Mars. Goodyear then used the discovery and produced a new radial tire with 10,000 mile tread life expectancy.

---

[25]Chase Econometric Associates, Inc. (1976)

For the Rosetta mission, a special camera was developed that was able to differentiate between different shades of gray of the comet. This technology is now used for early forest fire detection by observing forests and indicating smoke vapor or fog. These are just few examples of the technological spin off of space exploration.

Several analyses attempted to identify and measure that portion of economic growth attributable to NASA R&D expenditures. A Midwest Research Institute study[26] of the relationship between R&D expenditures and technology-induced increases in GNP indicated that each dollar spent on R&D returns an average of slightly over seven dollars in GNP over an eighteen-year period following the expenditure. The Space Division of Rockwell International conducted a study of the macroeconomic impact of NASA R&D programs. Rockwell investigated the relationship between NASA's Space Shuttle program and employment in the state of California. Using an econometric model developed at UCLA, Rockwell estimated that the Space Shuttle program generated an employment multiplier of 2.8; that is, direct Shuttle employment of 95,300 man-years in California produced an increase of 266,000 man-years in total employment.

Studies conducted in Europe show similar results. Norway examines its industry every year since the 1990s. Norway has detected a positive spin off factor derived from its space investments. For each million Norwegian kroner of funding through the national or ESA programs, the Norwegian space sector companies have on average attained additional turnover via new markets worth 4.7 million Norwegian kroner in 2013. Denmark found the same type of effect in its industry, with a positive 3.7 spin-off factor, whereas each million euros of Danish contributions to ESA have generated a turnover of 3.7 million euros.

3.5 New private space initiatives

Within the last decade, several private initiatives have surfaced promoting the expansion of the space economy beyond earth orbit. A partial list of these companies is shown in table 3.4.

[26]Midwest Research Institute (1971)

Table 3.4 – Private Space Initiatives

Companies	Mission
Lunar development companies	
Golden Spike Company	Golden Spike Company is planning affordable priced lunar orbital and surface expeditions.
Shackelton Energy	Shackleton intends to undertake lunar prospecting. They plan to establish a network of "refueling service stations" in low Earth orbit and on the Moon to process and provide fuel and consumables for commercial and governmental customers.
Moon Express	Moon Express plans to offer commercial lunar robotic transportation and data services with a long-term goal of mining the Moon.
Excalibur Almatz	**Excalibur Almaz** plans to implement lunar tourism, plus wider space services, and space-flight technology.
Asteroid development companies	
Planetary Resources	Planetary Resources' short-term mission is to develop low-cost robotic spacecraft to explore resource-rich asteroids. The long term mission is to develop the most efficient capabilities to deliver these resources directly to both space-based and terrestrial customers.
Deep Space Industries	DSPs mission is asteroid mining. DSP plans to offer space-based refueling, power, asteroid mining, and space manufacturing.

Credit: author

These companies are seeking the best path to commercial success and aggressively leading the way beyond Earth orbit. One short-term strategy would produce fuel, oxygen and water for fueling spaceships, existing satellites, and life-support systems. A medium-term strategy is centered on leveraging space minerals into a logical configuration of complexity and infrastructure needs. As humanity moves towards the planets of the solar system, the use of minerals found throughout the solar system will be instrumental to their success. The question is not "how" to leverage space-mineral resources, but "how best" to leverage them. Among their exploitation will be to use space minerals for survival, habitats, and space infrastructures during humanity's expansion into space, and when profitable for trade with Earth.

Investment in private space initiatives by members of the Forbes Billionaires is becoming increasingly fashionable. A list of six space-investing billionaires (as counted by Forbes) in 2011 has grown to ten in 2013. The joint net worth of these individuals is over 106 billion dollars as shown in Table 3.5. Today, other high net-worth individuals investing in space include: Robert Bigelow (Bigelow Aerospace); Charles Simonyi (Planetary Resources); Anousheh Ansari (X-Prize); Dennis Tito (Inspiration Mars); Naveen Jain (Moon Express); Barney Pell (Moon Express); Tom Pickens (SpaceHab); and John Carmack (Armadillo Aerospace).

Table 3.5 – List of Space Investing Billionaires

Rank	Name	Age	Net worth (billion dollars)	Source	Space investment
19	Jeff Bezos	49	25.2	Amazon	Blue Origin
20	Larry Page	40	22.8	Google	Google Lunar X Price
21	Sergey Brin	40	23.0	Google	Planetary Resources
53	Paul Allen	50	15.0	Microsoft	Spaceship One
138	Eric Schmidt	68	8.2	Google	Planetary Resources
272	Richard Branson	53	4.6	Virgin Group	Virgin Galactic
527	Elon Musk	42	2.7	PayPal	SpaceX
831	Guy Laliberte	53	1.8	Cirque du Soleil	Visitor to ISS
922	Rma Shriram	56	1.7	Google	Planetary Resources
1051	Ross Perot Jr.	54	1.4	Oil & Gas	Planetary Resources
Total			106.4		

Credit: author

3.6 The political economy of the old space age

This period saw the rapid growth worldwide of the satellite commercial sector. It marked the third stage of Schumpeter's innovation process, the diffusion stage, in which the new products and processes spread across the potential market. Governments continued to contribute to the growth of the space commercial sector through legislation and policies. Government favored the best possible uses of competition, both nationally and internationally, to provide means of guiding individual efforts in the space commercial sector. Nationally, competition policies adopted by various branch of governments helped the growth of a strong and growing space commercial sector. By establishing an international regulatory framework for the commercial satellite sector, governments designed a mechanism for competition among private companies and not among sovereign states. Although limited to the commercial satellite sector, these measures enhanced the peaceful utilization of space.

The satellite commercial sector created new markets, and new possibilities for economic growth through efficiencies that did not previously exist. The impact of new technology occurs at the diffusion stage and thus the measurement of impact is very much a measurement of how the economy changes as new technologies are introduced and used. In economic terms, a production function describes the way in which quantity of product varies if quantities of factors vary. Innovation combines factors in a new way as it varies the form of the function. Innovation is thus the setting up of a new production function. Through innovation, entrepreneurs move resources from old and obsolescence to new and more productive employments, which is the very essence of economics and most certainly of a modern economy. The entrepreneurial function does not consist in either inventing anything. It consists in getting things done in response to an economic situation or certain features of it.

Governments continued to invest in space activities. But, facing limited budgets, space agencies redirected their efforts to robotic space exploration of the solar system and away from human space exploration, which was limited to suborbital flights and short stays at the International Space Station. Robotic space exploration contributed to scientific advancement in the understanding of the solar system and the universe. Building and managing the Space Station has spread information among different countries participating the project, thereby facilitating the development of new technologies in many countries. It has also stimulated industrial and global collaboration. Through their programs of Earth observations, they contributed to a better understanding of climate change. Space agencies developed R&D programs and technologies that will weight heavily in the future. But space agencies still operate under national governments whose main goal is the technological dominance in the space sector. They collaborate with each other only when forced by the costs of some space project.

In summary, the growth of the space economy can be comprehended only through an understanding of the interactions among diverse economic agents and governments, and the extra systemic environment in which governments operate. Although it is useful to think of the extra systemic environment as being part of the landscape, it undergoes continuing evolution. With the end of the Cold War, and the integration of the world economy and its rapid growth, new challenges face humanity, from pressures from population growth on natural resources to climate change. Space technology could be part of their solution. It would involve exploiting the full power of the Sun and moving the space economy beyond Earth orbit. New entrepreneurial initiatives are emerging promoting these activities. What policy and institutional changes are needed will be explored in the next chapter.

Bibliography

Airbus: *"Growing Horizon: Global Market Forecast 2017-2036"*, Airbus Publishing, 2016.

Bignami G. et al.: *"Project 242: fission fragments direct heating for space propulsion - program, synthesis and applications to space exploration"*, Astro Acta, 2011.

Craig H. W.: *http://ntrs.nasa.gov/archive/nasa/casi.ntrs.nasa.gov/20050160960_2005161052.pdf*, NASA TM-2005-213559, 2005.

OECD: *"The Space Economy at Glance: 2014"*, OECD Publishing, http://dx.doi.org/10.1787/9789264217294-en

U.S. Government Accountability Office: *"The Intelligent Transpiration Systems"*, Report to the Committee on Science, Space, and Technology, House of Representatives, GAO-12-308, 2012.

The Space Foundation: *"The Space Report, 2015"*, The Space Foundation, 2015.

Chapter 4

The political economy of the new space age

"Change is the law of life. And those who look only to the past or the present are certain to miss the future."

John F. Kennedy

Pressures from population growth and limited resources are real today. It is expected that world population will continue to increase in the next fifty years. Today world's population is about seven billion. Although it is difficult to forecast the growth of the world population, most observers agree that the world population could hit ten to fifteen billion people in 2050. There is still a debate whether the forecast level of population on Earth will reach the threshold compatible with limited resources. But current studies indicate that, although physically mineral resources are abundant, critical minerals and fossil fuel resources may reach the limits of economic feasibility in the next twenty to thirty years. This may lead to intense rivalries among different powerful nations to control exiting resources, and to a more unstable world.

Space technology could help in avoiding this conundrum. The exploitation of off-world minerals, which are limited/critical on Earth, is creating an opportunity to enlarge the mineral resource base on Earth, and to lead to a more stable world. New entrepreneurial forces are already emerging and committed to the enlargement of the space economy beyond Earth orbit. These facts, together with the lowering of the costs of accessing space, have created the conditions for the advent of a new space age.

But new and extensive international cooperation between space agencies, and the creation of a system which makes as much use as possible of the spontaneous forces of society are needed for the new space age to come about. In this perspective, space agencies would provide initial support in exploration and science, in advancing critical technologies, and in building space infrastructure. The private sector would then take the lead in creating new markets and in expanding humanity's presence in space. But how the social, political, and economic pressures and interests influence the political process will determine the future course of events. In this contest, two important questions arise: why governments should dedicate significant public resources, above the present levels, to going into space; and whether present space institutions are adequate to support the development of space resources by private companies.

4.1 Why governments should dedicate significant public resources to going into space

To begin to understand the political economy of the new space age, one must first understand the reasons why governments should devote significant resources to going into space. A reply to this question depends on the economic and political benefits resulting from space programs and the uses of space resources.

4.1.1 Economic benefits

Economic benefits arise from technological innovations, and the efficiencies generated by private sector activities in space; and from consumer welfare effects measured by the benefits to consumers beyond the value they paid for those products and services.

Technological innovations

Economic effects arise from technological innovations induced by space programs and private sector activities in space. Technological innovations from space programs and private sector activities in space could start a leap forward in productivity and growth here on Earth. In the short term, downward pressure on launch prices and cost-saving advances in satellite technology will combine to open the door for small and mid-size space companies to enter the market, providing new niche services and solutions to a growing number of customers. Reduced costs of accessing space help in creating new markets and services. Already SpaceX calls for launching a constellation of 4,000 small and cheap satellites that would beam high-speed Internet signals to all parts of the globe, including its most remote regions. If successful, the attempt could transform SpaceX from a pure rocket company into a massive high-speed-Internet provider. It would compete with major companies in the developed world but also make first-time customers out of the billions of people who are currently not online.

As a possible result of some of the innovative trends, small satellites have become in the past five years very attractive, due to their lower costs and shorter production lead times. There is still a natural trade-off to be made between a satellite's size and its functionality: the smaller a satellite is, the fewer useful instruments it can carry, and the shorter its lifetime since it carries less fuel. But advances in both miniaturization, improved satellite integration technologies, and satellites with electric hybrid propulsion[1] have dramatically diminished the scope of that trade-off. In terms of future developments, research is needed in networked systems of distributed, cooperating small-satellites, away from the current traditional, large, multifunctional satellites. Some experts see this as an evolution similar to computers, i.e. large mainframe computers of the 1970s have evolved into networks of small computers connected via Internet.

Reduced orbital launching costs may make space-based solar power[2] (SBPS) economically workable, which would add an important source of clean energy to reduce global warming. Potential advantages of collecting solar energy in space include a higher collection rate and a longer collection period due to the lack of a diffusing atmosphere, and the possibility of placing a solar collector in a geosynchronous orbit where there is no night. But early architectures to accomplish this vision were technically complex and not economically viable. Some of the issues with these earlier solar power satellite concepts - particularly involving technical feasibility - were addressed by a new SBPS concept[3]. But the costs of launching material into orbit is still unsolved. According to J. C. Mankins, such projects would become economically feasible if the launching costs are reduced to 1,000 dollars per kilogram of payload.

A good indicator of the average minimum cost at which electricity must be sold in order to break-even over the lifetime of the project is the levelized cost of electricity. Table 4.1 report the levelized costs of different sources of electrical power, including the space based solar power, and the capacity factor, which is the unitless ratio of an actual electrical energy output over a given period of time to the maximum possible electrical energy output over the same amount of time. The levelized cost of electricity is an economic assessment of the average total cost to build and operate a power-generating asset

---

[1]Electric propulsion technologies are classified into three categories: electrothermal, electrostatic, and Plasma. These are types of propulsion that have been under study for more than thirty years in several countries, particularly the United States, France and the Russian Federation. The main advantage of electric propulsion used for commercial satellites is that due to the relatively lower weight of the satellite. An operator can embark more marketable capacity (i.e. transponders on board telecommunications satellite to lease to its customers), in place of the fuel the satellite would have needed if it used a classic chemical propulsion system.

[2]Space-based solar power is the concept of collecting solar power in outer space and distributing it to Earth. As considerable fraction of incoming solar energy (55–60%) is lost through the Earth's atmosphere by the effects of reflection and absorption, space-based solar power systems by converting sunlight to microwaves outside the atmosphere avoid these losses.

[3]John C. Mankins (2012)

over its lifetime divided by the total energy output of the asset over that lifetime. The capacity factor is a good indicator of the substitutability between renewable energies and fossil fuels. The lower is this factor the less the substitutability, because fuel may not be available.

Table 4.1 – Levelized costs of different power sources

Plant type	***Capacity factor in %***	***Total levelized costs in dollars/MWh***
Fossil fuels		
Conventional coal	85	95.6
Advanced coal	85	115.9
Advanced coal with CCS	85	147.4
Natural Gas conventional CC	87	66.3
Natural gas advanced CC	87	64.4
Natural gas advanced CC with CCS	87	91.3
Renewable energies		
Advance nuclear	90	96.1
Geothermal	92	47.9
Wind	34	80.3
Wind offshore	37	204.1
Solar photo voltaic	25	130.0
Solar Thermal	20	242.1
Hydro	52	84.5
Advanced SBSP	98	90.0

Credit: author. **Source:** US Energy Information Administration (2013), except for SBSP which was derive by J. C. Mankins (2012).

Among the renewable energies, nuclear, geothermal, and SBSP are highly substitutable with respect to fossil fuels. SBSP is highly competitive with fossil fuels, provided launch costs are substantially reduced. Space based solar power would reduce greenhouse gases emissions from human activities such as coal or wood burning, and use of diesel oil in power generation. It is evident that space based solar power will not solve the problem of global warming, which is centered on the stabilization of the atmospheric concentration of the greenhouse gasses. However, it would add to other clean energies to mitigate global temperature rising over a threshold, avoiding the materialization of a catastrophic scenario. The global warming catastrophic scenario arises because of its feedback on GDP, which could become very high if changes in global temperature reach a threshold over 2 °C.

In *the medium term,* mining the Moon and/or the near earth asteroids and building space infrastructure would require technological innovation in robotics, computer, transmissions, 3D printing, and mining equipment industries, which could affect the economies here on Earth. Technical innovation in robotics can have strong efficiency effects in industry. Reducing the cost of the energy component of production is on the mind of manufacturers all over the globe. There are several strategies to meet this challenge. One proven method is to deploy robotics. Robots with servo motors are already very efficient relative to power usage. Because robots can operate unsupervised, they can produce items while people are not present. The ability to operate in the dark or in unheated environments can lead to substantial energy savings on the part of manufacturers. Robots have improved control of motion, so there is a lot of energy savings because acceleration and deceleration are done in a much more controlled way. This improvement comes from more sophisticated algorithms in control technology that would be developed for space mining.

3D printing technology can receive a strong impulse by space infrastructure manufacturing. Building space infrastructure requires moving beyond printing basic supplies[4] and into printing full-scale components of a space station and spacecraft. At present, NASA is funding research of a robotic 3D printing and assembly system that could build structures larger than half a mile wide in orbit[5]. ESA and the European Commission have embarked on a project to

---

[4]Some construction industry, for example, uses it to create one-of-a-kind elements covering the joints of steel pillars supporting the roof. Creating metal casts to fulfill these functions would be very difficult, if not impossible, and therefore prohibitively expensive. In this case 3D printing lowered the cost of the building. Some engineering and robotic industries use 3D printing to create their own spare parts when needed, thereby reducing costs of acquiring and storing spare parts. Almost the entire automotive industry uses 3D printing. For example, Volvo used this technology to shorten the wait time for sets of metal tools used to install car seats by ninety four per cent. The result was a significant drop in costs. Ford, thanks to using 3D printing to create prototypes of Mustangs was able to shorten time needed by thirty times and costs by over hundred and fifty times.

[5]Tethers Unlimited recently received 500,000 dollars from NASA to continue developing Spiderfab, a robotic 3D printing and assembly system.

perfect the printing of space-quality metal components[6]. This project is focusing on serious engineering components made of very high-tech alloys, using lasers, electron beams and even plasma to melt them. Spillover effects on the economy here on Earth can be dramatic. Building high-quality, intricate shapes components would have massive cost savings in manufacturing industry, particularly automotive, engineering, and construction industries. For example, aiming towards zero waste production, a kilogram of titanium would go into a kilogram of the end product, which translates into enormous cost savings and energy efficiency.

Efficiencies generated by private sector activities in space

In the medium term, efficiencies generated by private sector activities in space derive mainly from space mining. Pressure from increasing world population and limited/critical mineral resources is already putting some strain industry. For at least three decades, scientists and economists[7] have debated whether humanity is rapidly depleting the mineral resources on which it depends. The prevailing opinion is that there is no immediate concern about the physical availability of mineral resources While we are not running out of any mineral, extraction is becoming more and more difficult as the easy ores are depleted.

Mineral experts[8] say that we are reaching the limits of economically feasible extraction of a number of mineral commodities, including metals and fossil fuels. The extraction of metals such as copper, zinc, nickel, gold and silver could peak and then decline in less than 20 years. Other mineral commodities that are irreplaceable in their industrial uses may become in short supply in the near future: platinum group metals for automotive catalytic converters, rare earths for magnets, and more. According to these studies, the world will never run out of those minerals, but extracting them will prove far more expensive due to deteriorating ore grades. That means that more rock has to be shifted and processed to refine and extract the desired raw material, requiring more energy and leaving more wastes. If this is occurring in parallel to the depletion in energy sources, more energy has to be used to extract a given quantity of energy and therefore to extract from a given quantity of ore, lifting mining costs.

---

[6]The AMAZE project – Additive Manufacturing Aiming Towards Zero Waste & Efficient Production of High-Tech Metal Products – involves 28 industrial partners across Europe.

[7]See Kesler S. (1994); and Craig, J. R. et al. (2001).

[8]U. Bardi (2011) and Club of Rome (2014)

Chart 4.1 report some selected base metals orc grade in exiting mines. It indicates that there is a negative trend in the quality of ore grades, which, if continues, would make the costs of extracting metals higher.

Chart 4.1 – Selected base metals ore grade in exiting mines

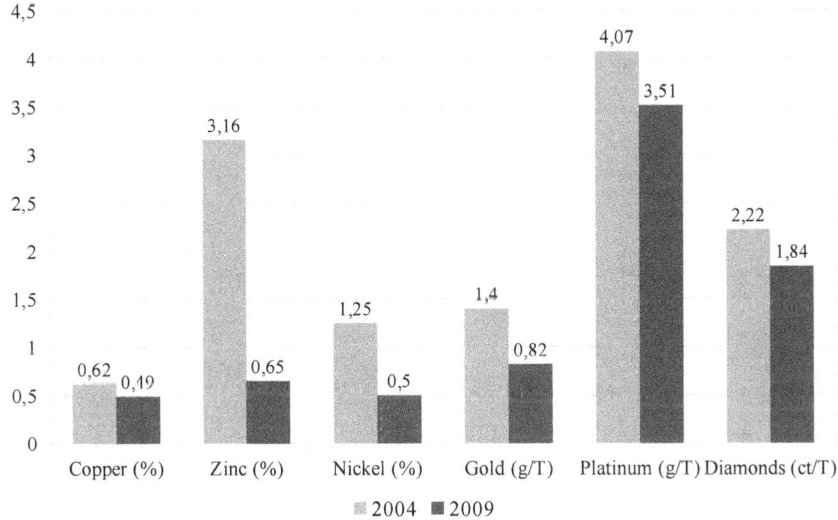

Credit: author. **Source data:** Statistica (2017)

Table 4.2 reports the energy intensity (EROEI) and other metrics for some critical minerals. Critical minerals are referred to as minerals whose supply is constrained while demand is high, and whose market is global in scope. On the supply side, criticality is caused by the confluence of various factors: physical constraints; the influence of non-primary market actors on the market, such as governments; and the fragility of the supply chain. This last one is caused by disruptions arising from operating dislocations. They can be either stochastic (e.g., natural disasters), organizational (e.g., labor unrest and government policies) or institutional (e.g., non competitive firms' behavior).

Table 4.2 – Metric for some critical minerals in 2016

	Ore grade (%)	Energy intensity EROEI	Prices in US$/kg	% produced in a single country
Copper	0.42	0.10	0.78	30
Nickel	0.5	0.195	9.32	23
Platinum (g/T)	3.51	200	34,616	70
Gold (g/T)	0.82	100	45,372	15
Rare earth	0.0037	n.a.	n.a.	93

Credit: author. **Sources:** Ernst W. G. (2000); Alonso et al. (2007); and USGS (2011). Note: Energy intensity is expressed in gigajoules per kilogram.

The first and second metrics in table 4.2 consider physical constraints determined by efforts needed to get each kilogram of metal. Two group of metals are particularly interesting for this analysis because of the interconnections with climate change: PGM and rare earth elements. Costs of extracting and refining PGM and rare earth elements are greater than those of other metals because of low ore concentrations and the high energy intensity. For example, platinum ore grade varies from 0.5 to three grams per metric tonne. This is three orders of magnitude less than the average grade of copper, nickel, tin, and zinc. The energy needed in PGM and rare elements mining and refining is around 1000 orders of magnitude greater than those for other metals.

The largest PGM deposits are in countries that are politically unstable or under the influence of non-primary market actors. Hence the fragility of the supply chain to disruptions from operating dislocations is high. The two largest producers of PGM are South Africa and Zimbabwe. In the past, labor unrest and energy shortages in South Africa were the main reasons behind the high PGM's price volatility. Zimbabwe passed legislation in 2010 requiring large corporations to have fifty one per cent domestic ownership in mining, causing delays in investments to expand existing operations and to explore new deposits. On the demand side, PGM demand derives mainly from the auto catalyst and jewelry industries, but they are also used in an extensive range of other products and applications. These include glass, electronics, non-transport emission-control equipment, and industrial manufacturing. Collectively they account for twenty five to thirty per cent of the total demand.

Looking ahead, analysts expect the PGM demand to grow at an average annual rate of six to ten per cent per year over the next twenty years. A major boost to global PGM demand will come from emerging countries, such as China and India, as environmental legislation to reduce auto emissions is expected in these countries over the next ten years. A common concern raised over PGMs is the extent of remaining resources. Based on 2009 production of about 450 tons and some 80,000 tons of certified resources, there are 175 years supply at present rates (average annual growth rate of 2.5 per cent between 2000 and 2016) and based on conventional mining and processing technology. But, if a six per cent growth in production is assumed, by 2050 annual production will have reached about 4910 tons with cumulative production about 80,000 tons. This forecast can be mitigated by recycling. But a more critical and pressing issue is the deteriorating ore grades of PGM. This is particularly true in Russia, where palladium ore grades are declining. In South Africa, at present, economic resources in the Bushveld are located between one to two kilometers depth. Increasing production would require going several kilometers deeper, with implications for the economic feasibility of the project. Alternatively, existing lower grade reserves in South Africa could be brought into production. Either way, it is reasonable to assume that by 2050 a potential crisis is looming, which would affect important parts of the modern industrial economy.

China has a near-monopoly in global production of rare earth elements. Although China produces around ninety seven per cent of the global supply, only fifty per cent of world rare earth elements reserves are actually in China. Rare earths mines in North America closed because of environmental contamination problems, while China appears to be willing to bear the environmental cost of extraction. Due to high domestic demand, the Chinese government cut export quotas in 2010 and 2012. This caused a rapid increase in prices. Unless technological innovations reduce environmental contamination, it is doubtful that rare earth mines in North America will be reopened. New sources are difficult to find. Despite their name, rare earth metals are fairly common in the Earth's crust, they are so widely scattered makes them difficult to mine. This means that China would retain its monopoly on the supply of rare earth elements.

Although exploration efforts, economic evaluations, and the search for solutions to environmental problems continue in North America and other countries, most experts[9] project that demand of rare earth minerals will outstrip net supply over the next twenty years, as rare earth elements are becoming increasingly integrated in new technologies, especially within the clean ener-

[9]National Research Council of the National Academies (2008).

gy, military, and consumer electronics sectors. While exact projections of demand of rare earth elements in future years are difficult to make, analysts have placed the lower and upper bounds for annual growth for total demand at five to nine per cent per year over the next twenty five years.

Demand will not grow uniformly for individual elements, but rather depends on growth in the markets for derivative products of the individual elements. The permanent magnets sector is expected to have large growth over the next century. This is mainly due to increasing demands for green energy. High-efficiency generators in turbines and electric motors require strong, yet light, permanent magnets. Currently, magnets using neodymium, praseodymium, and samarium (with dysprosium and terbium additives) are the strongest and lightest that are commercially available . By 2035, neodymium demand may be almost forty per cent of total demand compared to the twenty per cent of total demand that it is today. Assuming technology does not change significantly, constraints on the supply side may reach their peaks at the moment when demand for electrical motors and wind turbines may have reached their maximum due to the demand for clean energy.

Table 4.3 shows the demand for rare earth oxides, the main way rare earth elements are purchased. Permanent magnets, metal alloys, and catalysts generate the largest demand for rare earth elements, which in turn are driven by the demand for clean energy.

Table 4.3 – Demand for rare earth oxides in 2011 in metric tons

End use	
Permanent magnets	21,000
Metal alloys	21,000
Catalysts	20,000
Polishing powders	14,000
Phosphorous	8,000
Glass additives	8,000
Ceramics	7,000
Others	6,000
Total	105,000

Credit: author.

In the longer term, the supply of off-world PGM and rare earth elements would reduce the fragility of the supply chain and price volatility, both the large price fluctuations and/or restrictions on minerals' availability at any price. As PGM risk's source emanates from physical and political factors, reduced volatility may thus come from new supply sources free from political risk. Supply of off-world minerals is free from political risk, if accompanied by an international legal framework regulating the use of space resources. By breaking the monopoly of China, off world supply of rare earth elements would increase competition and stabilize prices. In summary, off world supply of critical minerals would increase the world economy's efficiency because the automotive and electrical motor industries are important parts of the modern industrial economy.

Will off-world minerals supply have welfare effects? The answer is yes. It is known from economic theory that consumers value price stability, as risk averse consumers prefer deterministic consumption streams to risky ones with the same mean. As PGM and rare earth elements enter as an input in many industrial and consumer products, raw materials price volatility can be reflected in prices of industrial and consumer products. Despite manufacturers may find new ways[10] to mitigate the expenses, some of these costs may pass to customers in the form of price increases, unless new sources of these minerals are found. But increasing complexities in the extraction of these resources means that geological availability on Earth may not be available. Off-world minerals supply would thus have welfare effects.

Finally, the potential benefits to space mining reach far beyond just profit, economic growth, and expanding Earth's resource base. Mining on Earth of these minerals can be highly destructive to natural habitats, while mining in space does not damage any natural habitats. The typical scheme for a PGM project is a mine, flotation concentrator, smelter and refinery. Major environmental costs of a PGM project are related to water, solid waste disposal, and energy and greenhouse emissions (Table 4.4). As South Africa is the major producer of PGM in the world, data in table 4.4 refers to South African projects.

---

[10]Manufacturers may move from today's linear economy, where they mine, manufacture, use and throw away, towards a more 'circular economy', where one industry's waste becomes another industry raw material. Collaboration with suppliers and customers can keep used products, components and materials in circulation, while new business models that rethink ownership can shift value along such a supply circle. For instance, some companies have even begun to pioneer new business models that enable them to retain ownership of the materials used in the products they sell. By following a systematic approach that focuses attention on resources throughout the value chain, manufacturers can exploit a broader landscape of opportunity than they initially thought possible.

Table 4.4 – Summary of environmental sustainability metrics by PGM mine/project

Mine/project	*GJ/kg of PGM*	*Water m^3/kg of PGM*	*t CO_2/kg of PGM*
Bafokeng-Ras.	116	235	29.1
Lebowa	164	385	39.4
Potgietersrust	201	277	31.1
Amandelbult	106	209	26.3
Rustenburg	132	229	33.2
Union	190	237	42.9
Northam	226	1,612	78.3
Average	162	225	40.0

Credit: Author.

The typical range for unit water consumption is 209 to 1,612 cubic meters per kilogram of PGM, with an average of 225 cubic meters per kilogram of PGM. Despite the increased reporting of water consumption by PGM projects, no data are available for impact on water resources and water recycling. This is critical since water quality is the primary factor in determining its potential use, recyclability, and its potential impact on the environment. The typical range for unit energy costs is presently 106 to 226 GJ per kilogram of PGM, with an average of 162 GJ per kilogram of PGM. Most projects show variable unit energy costs over time, probably related to the increasing depth of the mines. As South Africa's electricity supply is dominated by coal (89.7%) with a small proportion of hydro-electricity (5.0%), CO_2 emissions related to PGM projects are relatively high. The only project appearing to be outliers in all metrics is Northam, one of the deepest mines in South Africa.

Mining, refining, and recycling of rare earths have serious environmental consequences if not properly managed. Most of these environmental impacts are at local levels. The concentration of rare earths in the ore is very low, so they must be separated and purified, using hydro-metallurgical techniques and acid baths. A particular hazard is mildly radioactive slurry tailing resulting from the common occurrence of thorium and uranium in rare-earth element ores. Improper handling of these substances can result in extensive environmental damage. The foul waters of the tailings pond contain all sorts of toxic chemicals and radioactive elements such as thorium. It is the tailings that are of particular concern as they are full of small, fine particles that can be absorbed into the water and ground surrounding a particular mine.

4.1.2 Political benefits

Political benefits of space programs come first in terms of short term economic growth and employment, particularly employment opportunities for young people, since one way politicians measure the benefits of large programs is through the jobs and revenues generated within their constituencies. But major political benefits come in terms of world political stability, particularly in sub-Saharan Africa. Africa has experienced more civil wars since the 1960s than any other major region around the world. The African continent has always been a field of intense economic rivalries among different powerful states, which sought to have access to its natural resources. At the same time, Africa has been a source of strong transnational conflicts over borders and economic, racial and religious issues. For these reasons, it has experienced extremely violent and bloody civil strife. In several African countries, the abundance of strategic commodities - oil, diamonds and other strategically important minerals - fuels conflicts. Experiences of countries such as Angola, the Democratic Republic of Congo, Nigeria, Sierra Leon, and Sudan confirm this fact. But experience suggests that no single cause explains these phenomena.

The empirical literature has found complex reasons underlying the conflicts in those countries. These include the following factors: their capacity for coercion; the state of their infrastructure; the condition of their national identity; and social cohesion within them. Ethnic and/or religious divisions make a country more prone to civil war. But the collusion among the local predatory elite, foreign governments, and global companies often further stoke these conflicts. Since 2001, China and Western countries are engaged in geopolitical competition in Africa. Beijing has taken a series of initiatives aimed at securing strategic sources of raw materials in Sub-Saharan Africa. China uses its vast international reserves to extend loans in dollars to African countries. This policy helps gain access to Africa's vast raw material's wealth. It displaces the typical game control of the West through the World Bank and the International Monetary Fund. Loans have been made to finance infrastructure. China committed more than eight billion dollars alone to Nigeria, Angola and Mozambique in 2006. In comparison, the World Bank extended around two and a half billion dollars to sub-Saharan countries. Through such moves, China is displacing the Western powers in the game of diplomacy. But aid programs have brought corruption. The political elites, often conniving with transnational companies, pass along little of economic value to their populations.

China and Western powers interventions went beyond the standards of international aid as they enabled the struggles between various factions in African countries through the arms' trade. Arms went to factions seeking control of strategic minerals to enrich themselves and the armies they use to keep control. Evidence of the arms' trade can be readily found by looking at military spending in sub-Saharan countries. This spending has increased from around twelve billion dollars in 2001 to more than eighteen billion dollars in 2016[11]. Moreover, geopolitical competition for control of strategic minerals contributed to the increased military spending of China and Western powers. Several factors explain these increases: real or perceived threats; the surge of terrorism; armed conflict; policies to contribute to multilateral peacekeeping operations; and the natural-resources curse.

Space mining would allow Western countries and emerging ones, such as China and India, to diversify their supply sources. Expanding Earth's resource base would allow African countries to utilize their own mineral resources for the growth of their industrial base. Greater political stability may help reforms of political systems, including social, administrative, and legal reforms in sub-Saharan Africa necessary for their national economic development. They may help the population and not the interests of current predatory elites. These reforms and greater political stability would strengthen local markets and stimulate the entrepreneurial spirit of these populations. This is needed in Africa to speed up growth. Political stability, less corruption, and higher growth rates would help in reducing the attractiveness of terrorism, and should reduce pressures from the large migrations that we are witnessing today.

Table 4.5 – Military and Space Agencies Expenditures (in billions of dollars)

	2016	***Extrapolation over 20 years***	*%*
Total military Expenditures	*1,690.0*	*32,354*	*100.0*
Military Expenditures USA	611.1	14,786	45.7
Military Expenditures Other Countries	778.9	17,568	54.3
Space Agencies Expenditures	*73.7*	*1,474.0*	*100.0*
NASA	43.6	872.0	59.2
Russia	2.7	54.0	3.7
ESA	5.6	112.0	7.6

Credit: author. **Source of data:** Military expenditures: SIPRI: "Yearbook 2016", SIPRI Publications, 2016 .

[11]See Stockholm International Peace Research Institute (2016).

The above developments may result in a more peaceful international environment and in a gradual reduction in global defense spending. According to the 2016 Stockholm International Peace Research Institute report, total military spending worldwide was 1.7 trillion dollars in 2016 (Table 4.5). Assuming unchanged policies in a period of twenty years, cumulative military expenditures will be 32 trillion dollars. Under our scenario, accumulated military expenditures could well decline by five to ten per cent over a twenty year period. This means 1.6 to 3.2 trillion dollars.

This would allow a redistribution of public resources to the numerous problems affecting our societies, including some increases in space agencies' funding. Higher space agencies spending would offset the negative impacts of short-term decreases in military spending. Higher spending benefits companies that work for the military-industrial sector, such as the aerospace industry. Widening space exploration would have the support of both the companies and the people who work for them due to their direct interests in these programs.

In the medium term, the possibility of opening a new frontier in space will motivate and direct creative process to press beyond the limits of what is known and to build tools with which to understand elements of the unknown. It will stimulate the interest in science of the best and the brightest among young people because of the challenges of developing the technologies for space mining and exploration. This would include not only physical science, astronomy, and engineering, but also medicine and tele-medicine, genetic, biology and immunology, pharmaceutical science, and psychology to overcome health and psychological problems faced by explorers and colonists in long term space travel. Ongoing research both at the International Space Station and in laboratory[12] here on Earth are already contributing to the solution of these problems. All this would benefit Earth because of the needs to advance cancer treatments, and improving the efficiency of medicine through tele-medicine and robotic.

4.2 Whether present institutions are adequate to the uses of space resources

The development of space resources implies extensive collaboration among space agencies, and the presence of an international legal framework allowing competition among private companies. For decades, space science was the poster-child for international cooperation. Beginning in the 1990s, international cooperation in space activities intensified in all space disciplines. But major space faring nations have different attitudes with respect to international collaboration.

[12]See Horneck G. et al. (2003).

4.2.1 The United States

Historically, other than in space science missions, the United States has resisted relying on its partners but has welcomed contributions that enhance its core programs or missions. There are certain impediments to international collaboration that pose policy challenges. An onerous US exports control regime is one such policy challenge. Furthermore, there are conflicting national policies in relation to data, intellectual property rights and protecting the competitiveness of the national space industrial base which may impede international cooperation. Finally, changing economics and political situations during space projects, often needing over a decade of commitment, the lack of transparency or trust, in particular the intentions of some countries in relation to the peaceful use of outer space, are important policy challenges to international collaboration.

NASA experienced a change in this trend when Russia was invited to join the International Space Station (ISS) partnership in the early 1990s. Today the ISS is an example of a successful partnership with each partner reliant on the other. There is no question that the ISS has taught the United States, Russia, Europe and Japan to develop new technologies and has stimulated industrial and global collaboration. International cooperation gained further momentum during the Obama Administration. National Space Policy 2010 made international cooperation prominent in the United States space policy. Overall, the goals of the 2010 US National Space Policy were to: energize competitive domestic industries; expand international cooperation; strengthen stability in space; increase assurance and resilience of mission-essential functions; pursue human and robotic initiatives; and improve space-based Earth and solar observation. The 2010 policy changed the language from one that exclusively referred to US space assets to one that includes all nations having the right to freely use space without any harmful interference.

The 2010 policy places greater emphasis on international cooperation and building a global sense of responsibility for sustaining the space environment for use by all and for peaceful purposes through the enhancement of Transparency and Confidence Building Mechanisms. Further, the 2010 policy underlines the importance of building partnerships with the commercial space sector and also the need for greater space situational awareness in terms of the better monitoring of space debris and taking measures to avoid in-space collisions. With regard to the human space exploration program, the Obama Administration went ahead with the planned retirement of the Space Shuttle and committed the US to the international Space Station project to at least 2020.

The Obama Administration canceled the Bush vision for a return to the Moon and consequently also canceled the Constellation lunar transportation systems program that was being developed. The Obama policy has, instead, declared that an asteroid should be America's next destination for human exploration. To this end, approved NASA project to build a new space station in cislunar orbit as a first step to a longer term human exploration of Mars. The Obama Administration also put emphasis on nurturing the commercial sector to develop crew and cargo transportation systems to serve the International Space Station. The cancellation of the Constellation program and the increased reliance on the commercial sector have been very controversial especially in Congress. A compromise was reached and, as an insurance policy in case the commercial sector fails to produce the required space transportation vehicles, NASA is now developing a heavy lift launch vehicle and crew module that will be able to serve both the ISS and missions beyond Earth orbit.

In summary, the Obama administration space policy went a long way in creating the conditions for international cooperation in space, although some of the policy challenges, such as conflicting national policies in relation to data, intellectual property rights, and protecting the competitiveness of the national space industrial base still remained unanswered. The advent of the new administration brought substantial changes in the space policy of the United States. The Obama administrations set NASA on a path that hands over many activities in Earth orbit to private companies, theoretically freeing up the space agency to focus on more ambitious efforts, like getting people to Mars. It is yet unknown whether the new administration will continue these policies. But some doubt exists on how much international cooperation will be there, and who will be the United States partners. Moreover, the missions to explore the asteroids and to build a new space station are controversial both in the White House and in Capitol Hill, which may put these projects into jeopardy.

The present political and organizational set up raise a bigger question. As the cancellation of Constellation and the push to scrap the asteroid mission show, NASA is often pulled this way and that by the president and Congress. This is not an ideal situation for an agency that is trying to plan out a crewed Mars mission and other activities twenty or thirty years in the future. At present, the NASA administrator is nominated by and serves at the pleasure of the President. To shield the agency from the ebb and flow of partisan politics, some changes need to be brought in. One suggestion is to make the NASA chief to serve a fixed term, and/or to appoint him by a panel composed by representatives of the scientific community, civil governments agencies, and national security agencies. The goal of these reforms should be to make NASA able to plan in a consistent and efficient way its long term space exploration missions.

4.2.2 Europe

In 1975, a group of European countries created the European Space Agency, a body independent of the European Union. The foundation of ESA was built on international cooperation at regional level. The legal framework for a European space policy was provided in Article 189 of the EU Lisbon Treaty, which was signed in December 2007 and came into force in December 2009.

Article 189 provides in part as follows:

> "To promote scientific and technical progress, industrial competitiveness and the implementation of its policies, the Union shall draw up a European space policy. To this end, it may promote joint initiatives, support research and technological development and coordinate the efforts needed for the exploration and exploitation of space."

Since its beginning, ESA relies on cooperation with other space agencies, particularly NASA. But the changed circumstances in the 1990s (i.e., tough legal restrictions on information sharing by the United States military) led to decisions to depend more on itself and on cooperation with other space faring nations. Space is strategic for Europe's independence, job creation and competitiveness. Space activities create high-skilled jobs, innovation, new commercial opportunities, and improve citizens' well-being and security. This is why we need to reinforce European space policy to best exploit its social and economic opportunities for industry and SMEs. In order to achieve these goals, Europe needs to keep an independent access to space.

In 2003, the European Commission issued a White Paper titled *Space: A new European frontier for an expanding Union – An Action Plan for Implementing the European Space Policy.* This followed several years of debate in Europe about whether the EU (as against the European Space Agency and EU Member States) should take a lead role in shaping the European space policy. This debate is still going on. This section would add some considerations to this debate.

Several challenges face Europe. In the short term, reduced costs of accessing space and competition from other space faring nations risk marginalizing Europe in the launching business. Breaking the monopoly of Arianespace would require policies to increase competition as a decisive step towards the European space sector's independence. In the medium term, Europe should adopt a "planet hopping" approach in pursuing human space exploration. This would imply a strategy of partnership with private companies in pursuing these endeavors, and an increase in the amount of resources dedicated to space exploration.

Fifty years ago, European countries understood that they did not have a critical mass to operate alone in space and created the European Space Agency. In the last fifty years, ESA has succeeded in numerous space programs pushing the frontier of knowledge and the European space industry development. But it is doubtful an intergovernmental institution could break the monopoly in the heavy launcher sector. A European federal space agency would be better suited to design new policies to increase competition as a decisive step towards the European space sector's independence. It would be better suited to develop new technologies and to stimulate industrial collaboration among European space industries. It could also facilitate bringing together flight crews, multiple launch vehicles, operations, training, engineering, and facilities, communications networks, and the European scientific community.

Following the logic of the two speed Europe, only the states willing to pursue further political integration would become members of the new federal space agency. This institution should be open, and other states could become a member at a later date. Participating nations in the new federal space agency should abolish their own national space agencies, which will become regional offices of the new federal agency. Heads of regional offices would form the board of directors of the new institution. The board of directors would nominate the general manager, who should be approved by the European Parliament. This agency would have its own budget, approved by the European Parliament. An unresolved question is what to do with the actual intergovernmental agency (ESA). If, as expected, major European countries would form the core of the new federal space agency, the best solution would be to dissolve the intergovernmental agency, and replacing it with bilateral agreements with countries not willing to participate in the new federal agency.

The new federal space agency's main tasks would be to propose a federal space program and space policy guidelines to be approved by the European Parliament, and to coordinate and govern their implementation. A federal space program should focus on: building space infrastructure, pursuing human solar system exploration on a "planet hopping" approach, developing new propulsion systems and scientific research in space. Space policy guidelines should center on: increasing competition, and stimulating European space industries and private sector activities in space.

This new institution should also manage state properties in space infrastructure, and be responsible for negotiating international partnerships with other space faring nations. International partnerships will certainly be necessary in building space infrastructure and in human space exploration of the solar system. A federal agency with an augmented budget would have a better bargaining position in seeking partnerships with other space faring nations in developing human space exploration and the space economy beyond Earth orbit. In summary, this new federal institution should act as a competitor and partner with other space faring nations in the conquest of space.

4.2.3 China

China's space program was initially institutionalized under the People's Liberation Army (PLA). Through a series of government reforms in the 1990s, the China National Space Administration (CNSA) was created under the civilian Commission of Science, Technology and Industry for National Defense. But the PLA continues to play a role in China's overall space activities, managing both human civilian and military space efforts, while CNSA handles robotic science projects and international cooperation. China has an indigenous launch and space-asset capability. In October 2003, China became the third nation, after Russia and the US, to successfully launch humans into orbit.

This is seen as paving the way for an eventual Chinese space station in low Earth orbit by 2020, a modular design similar to the earlier Russian Mir space station. China successfully placed spacecraft in the lunar orbit on 1 October 2010, and also successfully launched its second lunar probe Chang'e-2, which will lead to a lunar sample return mission possibly in 2017. This seems as a prelude to a human lunar landing. But China has not had a firm plan for a human landing on the Moon even though studies are reportedly underway. A White Paper, published in late December 2011, mentioned a human mission to the Moon for the first time but only in the context of studies on the preliminary plan for a human lunar landing.

Chinese efforts in outer space have been shaped by political, bureaucratic, and technical realities. In particular, the political turmoil of the Mao era and subsequent efforts to restore stability to the Chinese economy deferred many of their objectives. The process of setting and implementing national policy objectives in China is established mainly through White Papers released by the State Council. The latest Chinese government White Paper on China's Space Activities was published in December 2011 and states that the purposes of China's space industry are to: explore outer space and to enhance understanding of the Earth and the cosmos; utilize outer space for peaceful purposes, promote human civilization and social progress, and to benefit the whole of mankind; meet the demands of economic, scientific, and technological developments, national security and social progress; and improve the scientific and cultural knowledge of the Chinese people, protect China's national rights and interests, and build up its national comprehensive strength.

The idea of the space sector contributing to China's growth and to "protect China's national rights and interests" may appear subtle, but an analysis of the 2011 policy paper reveals a potentially troubling trend towards a space program with a more serious national security element. China's space activities and intentions are not transparent particularly in view of the dual-use nature of most space technology. Any effort to interpret Chinese space policy and decision making will be fraught with uncertainty.

There is an ongoing debate on the role played by the PLA in running the Chinese space program. Informed sources provide the following explanation: When the Standing Committee of the Politburo gave the go ahead for the human spaceflight program, the Party Central Committee created a new office under the Special Committee called the Human Spaceflight Project Office responsible for coordinating and implementing the program. At the policy level, space program decisions are made by the civilian leadership. However, the Chinese military is probably the most capable institution and thus, is best placed to provide the day-to-day management of the human spaceflight program. The dual use nature of Chinese space technology is seen as a major impediment to China entering in future international cooperation efforts to human space exploration.

4.2.4 Japan

Japan's civil space program has undergone several structural changes during the past five decades. After the reorganization in 2003 and the formation of the Japan Aerospace Exploration Agency (JAXA), the management structure of the Japanese space program, along with policy formulation and Japan's definition of what constitutes peaceful use of space underwent a significant change with the enactment of the Japan Basic Space Law in August 2008 and the Basic Plan for Space Policy released in June 2009.

The Basic Space Law of 2008 established a Strategic Headquarters for Space Policy. The government elevated the importance of the Strategic Headquarters for Space Policy by announcing that its Chair would be the Prime Minister with Acting Chairs being the Chief Cabinet Secretary and the Minister of Space Development. The Basic Space Law also provides that space will be used for the purpose of "*improving and reinforcement of information gathering functions and enhancement of warning and surveillance activities in light of the international situation, especially the circumstances in North East Asia*". The new space orientation is in recognition of the changing nature of security threats and challenges, including growing Chinese space capabilities. Reflecting this, the new policy is unambiguous in laying out the security emphasis, saying: "*We will engage in space development to directly utilize it for our nation's diplomatic and security policies, as well as for the Self-Defense Forces.*" This is a major shift in Japanese policy eradicating the long-standing restrictions on the involvement of the Ministry of Defense in space activities, established in 1969 through the Diet Resolution Concerning the Peaceful

Utilization of Space that severely restricted the utilization of space for national security purposes.

The Basic Plan for Space Policy was released by the Strategic Headquarters for Space Policy in June 2009. It is a five-year plan with a ten year horizon. The Plan articulates the following "Six Basic Pillars for Japan's use and R&D of space": (i) improve the national space-based infrastructure for more effective weather forecasting, telecommunications, food and energy management, positioning and navigation services etc.; (ii) contribute to enhancement of security by strengthening the national information gathering capability; (iii) promote the utilization of space for diplomacy by providing access to satellite imagery to Asian neighbors for disaster prediction and management, climate change research and other global environmental concerns; (iv) create an energetic future by promoting R&D of the following areas: build on Japan's space science successes and world-class expertise such as Moon and asteroid science missions, human space activity (ISS and the future international space station in lunar orbit), and a space solar power program; (v) foster strategic industries for the 21st century, by placing the space industry among strategic industries in the 21st century and enhance competitiveness by promoting space machinery smaller, serialized, communed and standardized"; and (vi) consider the space environment by take measures regarding the space debris issue.

Despite a more aggressive policy on security issues, the space policy of Japan relies on international cooperation with allies and friendly nations. JAXA has established cooperative programs with ESA, NASA, and the Indian Space Research Organization (ISRO). Japan international cooperation focuses on the following areas: (i) space science and Earth observation, including meteorology, and in Earth science to address global environmental and climate challenges; (ii) inter-operability of critical space systems, focusing on: space-based positioning, navigation, and timing; enhanced space situational awareness; use of space for maritime domain awareness; research and development in space technologies; and use of hosted payloads; (iii) continued utilization of the International Space Station, and ensure close cooperation for the next International Space Station in lunar orbit; and (iv) support international efforts to develop transparency and confidence-building measures to encourage responsible actions in and the peaceful use of space, such as an International Code of Conduct for Outer Space Activities.

4.2.5 Russia

After the demise of the Soviet Union, the Russian government established a central agency (Roscosmos) in charge of space science program and general aerospace research. In the early years, the agency suffered from lack of authority as the powerful design bureaus fought to protect their own spheres of operation and to survive. The 1990s saw serious financial problems because of decreased cash flow, which encouraged Roscosmos to improvise and seek other ways to keep space programs running. This resulted in Roscosmos' leading role in commercial satellite launches and space tourism. While scientific missions, such as interplanetary probes or astronomy missions during these years played a very small role, Roscosmos managed to operate the space station Mir well past its planned lifespan, contribute to the International Space Station, and continue to fly additional Soyuz and Progress missions.

The situation changed drastically from 2005 onward. As the Russian economy boomed throughout 2005 from high prices for exports, such as oil and gas, the outlook for funding Roscosmos appeared more favorable. The budget for 2006 showed an increase of thirty three per cent. Under the current 10-year budget approved, the budget of the Space Agency shall increase five to 10 per cent per year, providing the space agency with a constant influx of money. The last Russian strategy paper released in 2005 was titled "Major provisions of the Russian Federal Space Program for 2006-2015".

The stated principal objectives of the space program in 2005 were as follows: (i) development, accomplishment, augmentation and maintenance of satellite constellations to provide socioeconomic benefits, security broadcasting, relay, Earth remote sensing, ecological monitoring, emergency management, and fundamental space and micro gravity research; (ii) deployment and maintenance of the ISS Russian Segment elements for fundamental and applied research; (iii) support of the Russian Segment of the KOSPAS-SARSAT International Search and Rescue System; (iv) maintenance and enhancement of the Baikonur space port; (v) sustaining development of competitive rocket and space technologies.

So far, only a very few of the objectives and milestones have been achieved and a string of failures in six consecutive launches forced a change of leadership for Roscosmos in the spring of 2011. On December 2011, the Russian Prime Minister Vladimir Putin put Deputy Prime Minister Dmitry Rogozin and the Russian defense industry in charge of the Russian space program. Given that previously, Roscosmos always had direct access to the Kremlin, the announcement that Roscosmos director Vladimir Popovkin will now report directly to the Deputy Prime Minister is a reflection of the current state of the venerable but now troubled Russian space program.

Up to now, the Russia Space Agency has continued its collaboration in the International Space Station. Roscosmos is a major provider of launch services to other countries, in addition to launching satellites for other countries. Roscosmos is now looking ahead to a major Mars mission, ExoMars, which it is planing with the European Space Agency. ExoMars' first leg (the Trace Gas Orbiter) launched successfully in 2016, while a rover was delayed by two years until an expected launch in 2020. Russia is also interested in developing a series of robotic Moon missions, called Luna-Glob. But, until recently, budgetary restraints have put this program on hold. Recently, this program has been revived with the backing of Europe. ESA has announced it will contribute a drill and other instruments to Roscosmos's Luna 27 lander, an uncrewed craft set to touch down on the Moon in 2020. The new head of ESA, Johann-Dietrich Wörner, has called for a collaboration to make a settlement on the Moon, modeled after the International Space Station. If this first mission is a success, it could be followed by lunar rovers and a mission to bring samples of moon rock back to Earth.

4.2.6 India

Today, India's primarily civil space program is administered by the Department of Space (DoS), which was established in June 1972. The Indian Space Research Organization (ISRO), the Indian national space agency, is the primary research and development arm of the DoS, and other agencies of the DoS include the Physical Research Laboratory, the National Atmospheric Research Laboratory, the North Eastern Space Applications Centre, and the Semi-Conductor Laboratory. The Indian Space program under the DoS aims to promote the application of space science and technology for the socio-economic benefit of the country.

From the beginning, India was keen on developing its own launch capability so that its space program could operate autonomously. It has now the capability (GSLV-3) to launch satellites in geosynchronous orbit. To date, the PSLV has placed twenty five Indian and twenty seven foreign satellites into orbit on a commercial basis. The inherent dual use nature of satellite technology could be adapted for military purposes as well. Indeed, the dependence of the Indian armed forces on space technology is an undisputed fact. Indian analysts are of the opinion that presently, this use is limited to areas like communication and navigation, but in the future, the dependence is likely to increase for surveillance and even for counter-space capabilities. The GSLV-3 will also enable India to pursue its lunar science mission program. Due to a strong interest within the Indian scientific community, ISRO has begun implementing dedicated science missions. In October 2008, India joined the ranks of

China and Japan by launching Chandrayaan-1[13], the first in a series of lunar orbiters. In 2013, it launched the Mars Orbiter Mission (MOM). As MOM successfully entered Mars orbit on 24 September 2014, India became the first Asian country to do so.

India often emphasizes its strategy of international collaboration with leading space powers and has strengthened its relationships with Europe in its efforts to commercialize its launch capability. India has also worked with Russia to develop its own geostationary launch vehicle. India has extensive collaboration with the United States. ISRO collaborates with NASA's NISAR to develop a satellite (NASA-ISRO Synthetic Aperture Radar) that can help in providing crucial information on the biomass estimation and agriculture production estimation. The NISAR will also help in studying earth surface deformation, which could provide valuable inputs for future earthquake precursors.

4.2.7 The prospects of international collaboration in space

The above paragraphs indicate to most of the countries under analysis, with the possible exception of China, are engaged in one form or another of international collaboration in space exploration. However, all space faring nations' space policies are driven and conducted independently at the national scale. Looking at the future of space exploration and uses of space resources, the continuation of this paradigm is bound in its aggregate effects to be harmful from a purely economic point of view and, in addition, may generate international frictions in space.

We have already noticed that conflicting national policies in relation to data, intellectual property rights, security, and protecting the competitiveness of the national space industrial base poses some limits to international cooperation in space. These are mainly mercantilist arguments, including infant industry argument, spillover effects, national security, unequal exchange perspective, fair trade and state interests divergent from global income maximization. Mercantilist explanations are based on the underlying belief that state interests do not coincide with global income maximization, but that another set of interests are at play which forms a trade off with the latter. Despite being strongly and lengthily criticized in economic literature as sub-optimal, those protectionist policies are widely adopted.

---

[13]Chandrayaan-1 is a complex spacecraft with 11 instruments including instruments from Europe (ESA), US (NASA) and Bulgaria. The mission, designed to measure the chemical and mineralogical properties of the Moon, particularly the southern pole, enjoyed a huge success, and was instrumental in the discovery of water on the Moon.

If the paradigm of national interest prevails in the future, there is little hope of international order in space so long as every country would be free to employ whatever measures desirable to its own national interest, no matter damaging it may be to other countries. If space resources are treated as exclusive properties of these nations, if international economic relations in space are relations among nations instead of being relations among individuals, then they will become a source of frictions among nations, not only in space but also here on Earth. History tells us that it is an illusion to substitute competition for markets and raw materials among individuals with negotiations among nations. It would only transfer decisions made among individuals without recourse to force and only subject to the rule of law to powerful and armed states, subject to no superior law.

The debate on a system conducting to peaceful relationships in space continues. A central assumption of international policies based on national interests holds that anarchy is a defined condition of the international system, and that statecraft, and space policies are part of it, is the only way to ensure national survival and the pursuit of national interests. The paradigm behind these policies is that international organizations and other trans-state or sub-state actors hold little real influence in the face of states as unitary actors looking after themselves. On the other side of the debate, there are those claiming the need of an international political authority, which, without power to direct individuals to do what they want to do, is able to restrain them from actions that damage others. In this line of thought, many claim essentially that the powers of the international authority are strictly circumscribed to the rule of law.

Some will always say that the first line of thought is *politics as it is* while the second one is an example of *politics idealized*. However, the significance of both lies in their capacity to explain opposite phenomena, and, though both are clearly antithetical, perhaps the answer to the question of how the world operates lies not in the thesis and antithesis, but in the synthesis of both. One pragmatic approach is for state advancement blended with a belief in humanity's inherent potentials.

In this line of thought, future space exploration and evolution of the space economy foresee a primary role of individuals and private companies in investing and developing space mineral resources and in creating new markets, while the role of space agencies would be limited to advance science and technology for space exploration, and in building space infrastructure. Because of the large public investments in space exploration and in building space infrastructure, international cooperation among space faring nations would follow. Examples of international cooperation, in which all partners rely on each other, already exists (International Space Station) on which to draw upon.

4.3 International legal framework

Competition among private companies in developing space mineral resources without recourse to force will require an international legal framework. Also in this case, there exist examples of international legal frameworks on which to draw upon. There are two aspects of the international legal framework relevant to the discussion here: (1) ensure the use of the Moon and other celestial bodies to peaceful purposes; and (2) define property rights and use of space mineral resources. The first aspect is covered by the Outer Space Treaty[14], formally the *Treaty on Principles Governing the Activities of States in the Exploration and Use of Outer Space, including the Moon and Other Celestial Bodies.* The treaty entered into force on 10 October 1967. As of July 2017, 107 countries are parties to the treaty, while another 23 have signed the treaty but have not completed ratification.

The Outer Space Treaty represents the basic legal framework of international space law. Under the terms of the "Outer Space Treaty", space, including the Moon and other celestial bodies, cannot be appropriated through claims of sovereignty by any nation. However, the State that launches a space object retains jurisdiction and control over that object. Article VI of the Outer Space Treaty deals with international responsibility, stating that *"the activities of non-governmental entities in outer space, including the Moon and other celestial bodies, shall require authorization and continuing supervision by the appropriate State Party to the Treaty"* and that States Parties shall bear international responsibility for national space activities whether carried out by governmental or non-governmental entities. The State is also liable for damages caused by their space object. Among its principles, the Outer Space Treaty bars states party to the treaty from placing weapons of mass destruction in orbit of Earth, installing them on the Moon or any other celestial body, or otherwise stationing them in outer space. It expressly prohibits their use for testing weapons of any kind, conducting military maneuvers, or establishing military bases, installations, and fortifications (Article IV).

Definition of property rights in the Outer Space Treaty is more ambiguous. A clause of the treaty states: *"Outer space, including the Moon and other celestial bodies, shall be free for exploration and use by all States."* This clause falls short of defining property rights, although experts argue that, if space mining is considered to be celestial bodies' use, any material extracted becomes the

---

[14]See United Nations (1967)

mining company's property. Property rights are very important to mining investors. But the right to exclusively mine an ore body in space has a higher priority. Note that until celestial ore bodies, being it an asteroid or the Moon, are given property rights, they will be classified as intangible assets, and as such would be subject to impairment testing under International Accounting Standard (IAS). Uncertainty about property rights and the uses of resources could hinder the funding of these activities by increasing the risk premium. Private companies and financial institutions are not likely to risk their capital and the considerable effort to develop the mineral resources in space, if significant legal or political difficulties can disrupt the supply of these assets. So it stands to reason that uncertainty over property rights and space resources' uses could hinder development of these activities by the private sector.

The interpretation of property rights is thus important and can be done by national legal institutions.

The US Commercial Space Launch Competitiveness Act 2015 is intended to clarify these issues since the United States has obligations to regulate its own private space activities under international law. At the end of November 2015, President Obama signed legislation allowing commercial extraction of minerals and other materials, including water and volatile, from the asteroids and the Moon. The Act purports to create property rights over resources extracted from asteroids. It states: *"Any resources obtained in outer space from an asteroid are the property of the entity that obtained such resources, which shall be entitled to all property rights thereto, consistent with applicable provisions of Federal law."* The Act defines this entity as *"a person or company providing exploration or utilization services that are either organized under US law or subject to US jurisdiction".*

Recently Luxembourg passed a law establishing property rights for space resources. Luxembourg's law mirrors the U.S.'s Commercial Space Launch Competitiveness Act of 2015, which served to "promote the right of United States citizens to engage in commercial exploration for and commercial recovery of space resources," including water and minerals. Both the Luxembourg and U.S. laws make sure to respect the Outer Space Treaty of 1967. By mining celestial bodies, companies cannot lay claim to that body but only to the resources the company extracts from it. The Luxembourg law states that companies may only explore and use space resources if they obtain written mission authorization from the ministers of the economy and of space activities. The law is part of the country initiative to promote international cooperation in order to progress on a future governance scheme and a global regulatory framework of space resources utilization.

This US Act received some criticism, which can be extended to the Luxembourg law. Joanne Gabrynowicz[15], a space lawyer and editor emeritus of the Journal of Space Law at the University of Mississippi School of Law, argues that the US Act fails to address basic issues, such as who would license and regulate space mining operations, as well as larger issues, such as the legality of space mining operations under international law. She argues that current efforts to clarify the legal status of space mined resources, if approached the wrong way, could guarantee other nations to challenge unilateral actions and likely cause political competition among nations in space. The reverse is also a concern. Disagreements over space could influence disputes on Earth. In her view, failure to make sure other signatory nations are on the same page could lead to negative geopolitical consequences. If other nations interpret the Act as the United States playing loose with the Outer Space Treaty, they might decide to do so themselves on Earth in unpredictable ways.

Others in the space law community disagree with Gabrynowicz. Brian Weeden[16] is a technical adviser at the Secure World Foundation, a think tank focusing on strategies for peaceful uses of space. He does not expect the sort of conflicts we see over resources on Earth to take place over resources in space, partly because there are vastly larger quantities of such valuables in asteroids than have ever been mined from our planet. He also believes that the Outer Space Treaty is already sufficient legal justification for space mining and that waiting for an international consensus could drag on interminably. As space mining progresses, the United States should engage in a dialogue with other space actors and the international community. If there are issues illuminating particular shortcomings or holes in the international legal framework, they should be addressed then. Weeden's position can be shared, because burdening space mining company with extensive regulations can be counterproductive at the beginning. A process where the United States and other countries put the regulatory framework in place in stages to match the staged development of the industry is more sensible.

4.4 Conclusion

With the end of the Cold War and the demise of the Soviet Union, some claimed the end of history. But this claim does not match reality. Efforts to harmonize political and social structures around the world have been complicated by trends that were scarcely evident few generations ago. The most important of these trends is the surge in world population and the demographic imbalances between rich and poor countries. Population explosion

[15]See Joanne Gabrynowicz (2006)
[16]See Brian Weeden (2016)

has put pressures on the environment and mineral resources. Demographic imbalances resulted in population surge in one part of the globe (poor countries) and a technological explosion in the other part. The coming together of these trends is not a good recipe for a stable international order. Making the human race interplanetary is part of the solution to those problems. It would help in enlarging Earth's mineral resource base, proving new sources of clean energy, stimulating technological innovations that may help transitioning to a new industrial revolution, and creating the conditions for a more stable world.

Key to the success of space activities is the recognition of the distinct roles of governments and individuals in developing the space economy beyond Earth orbit, specifically the recognition of the role of private companies in developing space resources. Contrary to widespread believes, this is not a return to the nineteen century ideas or the principle of laissez-faire. There is all the difference between creating a system which makes as much use as possible of the spontaneous forces of society, and the one based on the principle of laissez-faire. The first call for political powers, not mixed up in the economic game, to shape the vision for the future and to prioritize public spending accordingly, collaborate with other nations whenever it is in the interest of all, and design rules and regulations to ensure a fair competition among private companies.

The United States, as the dominant military, technological, and economic power, is in a position to exert a strong influence on these processes. If the United States were unwilling and/or unable to lead, the European Union should take up the torch. Since its foundation the European Union foreign and security policy seeks to preserve peace and strengthen international security; promote international cooperation; develop and consolidate democracy, the rule of law, and the respect for human rights and fundamental freedoms. European space technology has advanced during the last fifty years thanks to the space exploration programs of the European Space Agency, and the European Union policies to develop a competitive, independent and global European space industry.

The announcement by the United States and/or the European Union of their intention (i) to proceed with space exploration, in collaboration with other nations, leaving to private space companies to invest and to develop space resources, and (ii) to collaborate with other nations in the establishment of an international legal framework for the use and extraction of space mineral resources would give important messages to policy makers, emerging private space companies and financial markets, and other space faring nations. The most important message is that space activities will return economic, technological, and quality-of-life benefits on Earth. A second important message is that pushing the human presence and economic activities beyond Earth orbit opens a new frontier for the benefit of all humanity.

The financial resources for the realization of these programs could be found initially through the collaboration between governments and the private sector in developing off world mineral resources. A following more stable international political environment would provide the resources for the human exploration of Mars and its eventual colonization. The history of voyages of exploration and discovery of the modern period teaches that the strategic interests of governments combined with the interests of individuals in search of knowledge and adventure; and of business looking for new sources of raw materials, and new outlets for their products were instrumental to subsequent migrations. Voyages of exploration preceded migrations by creating the knowledge and the infrastructure that allowed future settlements. Our own past experience creates a precedent, or paraphrasing the poetic expression of Mark Twain *"history does not repeat itself but it rhymes."*

Bibliography

Alonso et al.: *"Material Availability and the Supply Chain: Risks, Effects, and Responses"* Environmental Science & Technology, vol. 41, pp. 6649-6656, 2007.

Bardi U.: *"The Limits to Growth Revisited"*, Springer, 2011.

Bardi U.: *"Extracted: How the Quest for Mineral Wealth is Plundering the Planet"*, Chelsea Green Publishing, 2014.

Chase Econometric Associates, Inc.: *"The Economic Impact of NASA R&D Spending: Preliminary Executive Summary."* NASA-2741, April 1975.

Craig, J.R. et al.: *"Resources of the Earth: Origin, Use, and Environmental Impact"* Prentice Hall Inc., Upper Saddle River, New Jersey, 2001.

Ernst W. G.: *"Earth systems: processes and issues"*, Cambridge and New York, Cambridge University Press, 2000.

Gabrynowicz J. I.: *"The Outer Space Treaty and Enhancing Space Security"*, UN Institute for Disarmament Research, 2006.

Horneck, G. et al.: *"HUMEX. A Study on the Survivability and Adaptation of Humans to Long-Duration Exploratory Missions"*, Noordwijk, 2003.

Kesler S.: *"Mineral Resources, Economics, and the Environment"* Macmillan, New York, 1994; and Craig, J. R. et al.: "Resources of the Earth" Prentice–Hall, Englewood Cliffs, NJ, 2001.

Klinenberg E.: *"Going Solo: The Extraordinary Rise and Surprising Appeal of Living Alone"*, Penguin 2012.

Midwest Research Institute: *"Economic Impact of Stimulated Technological Activity."* Kansas City, Missouri: Midwest Research Institute, November 1971.

National Research Council of the National Academies: *"Minerals, Critical Minerals, and the U.S. Economy"*, The National Academies Press, Washington, D.C., 2008.

Stockholm International Peace Research Institute: *"SIPRI Military Expenditure Data Base"*, 2016

United Nations: *"Treaty on Principles Governing the Activities of States in the Exploration and Use of Outer Space, Including the Moon and other Celestial Bodies"*, http://disarmament.un.org/treaties/t/outer_space

USGS: *"Mineral Commodities Summary, 2011"*, U.S. Government Printing Office. Washington D.C., 2011.

Weeden B.: *"Outer Space: Ungoverned or Lacking Effective Governance?"*, New Approaches to Managing Human Activities in Space, The SAIS Review of International Affairs, Vol. 36, No. 2, Dec. 2016.

Chapter 5

The space economy beyond Earth orbit: space mining

"I think we are at the dawn of a new era in commercial space exploration."

Elon Musk

Over the past decades the number of public and private actors involved in space activities worldwide has increased. The space economy includes many countries and commercial activities. Several mature downstream activities have reached mass markets. At present, all commercial space activities are carried out in the vicinity of Earth. But, within the last decade, a number of private initiatives have surfaced promoting deep space exploration and the exploitation of space mineral resources. Three factors are behind the emergence of these companies: downward pressure on launch prices, space agencies' exploration programs, and global problems. This chapter is focusing first on the primary "driving forces" at work in the present and their dynamics shaping the future. Second, it analyzes the geology of the Moon and the near earth asteroids. A clear understanding of these mineral resources is necessary to evaluate the potential of space mining. Third, it examines the economics of these projects.

5.1 Low costs of accessing space

In chapter 3, we have seen that costs of accessing space are coming down. Adaptation of new industrial qualification procedures is being pursued to mass produce spacecraft and launchers, based on existing experience and data from high volume industries, which would induce lower costs of accessing space. This process has been promoted by SpaceX. The business model is based on vertical industrial processes (i.e. more than 70% of each Falcon launch vehicle is manufactured at the SpaceX production facility) and mass production, inspired by the automobile sector and not used before in the space industry. The company's production volumes are constantly increasing, with production growing more than five times year-over-year - 2 Falcon rocket cores produced in 2012, 3 in 2013, and 17 to be produced in 2014 and 2015.

Competition is producing pressure on other aerospace companies in the United States, in Europe, China, and India to review their technologies and reduce costs. Lower costs of accessing space and technological innovations will help the evolution of the space economy beyond Earth orbit.

But how to improve costs to the Moon and beyond is tricky. It would require not only reducing costs of escaping Earth gravity, but also full re-usability of the rockets, and refilling them in orbit to maximize the payload of the rockets. The development of Falcon Heavy and its tests in late 2017 respond to the first requisite, i.e. reducing costs of escaping Earth gravity. The Falcon Heavy falls into the super heavy-lift range of launch systems under the classification system used by a NASA human spaceflight review panel. The initial concept envisioned fifty three tons payload to low earth orbit, 16 tons to trans lunar trajectory, and fourteen tons on a trans-Martian orbit to Mars. These maximum payloads are achieved when the rocket flies a fully expendable launch profile, not recovering any of the three first-stage boosters.

SpaceX intends to recuperate the first stage of the rocket. All three cores of the Falcon Heavy arrange the engines in a structural form SpaceX calls Octaweb, aimed at streamlining the manufacturing process, and each core will include four extensible landing legs. To control the descent of the boosters and center core through the atmosphere, SpaceX uses small grid fins which deploy from the vehicle after separation. After the side boosters separate, the center engine in each will burn for a few seconds in order to control the booster's trajectory safely away from the rocket. The legs will then deploy as the boosters turn back to Earth, landing each softly on the ground. Both the grid fins and the landing legs on the Falcon Heavy are currently undergoing testing on the Falcon 9 launch vehicle, which is currently used for vertical landing.

Under the competition of SpaceX, other companies and space agencies, both in the United States and other space faring nations, are looking to way to reduce costs. NASA is developing its own Space Launch System (SLS). The SLS is to be the most powerful rocket ever built with a total thrust greater than that of the Saturn V, putting the SLS into the super heavy-lift launch vehicle class of rockets. The SLS launch vehicle is to be upgraded over time with more powerful versions. Its initial Block 1 version is to lift a payload of seventy metric tons to low Earth orbit (LEO). Block 2 will replace the initial Shuttle-derived boosters with advanced boosters and is planned to have an LEO capability of more than 130 metric tons. These upgrades will allow the SLS to lift astronauts and hardware to destinations beyond LEO: i.e. on a circumlunar trajectory. The SLS will launch the Orion Crew and Service Module and may support trips to the International Space Station if necessary.

The SLS costs are undoubtedly higher than the ones of Falcon Heavy. But the Falcon 9 Heavy's payloads to trans lunar and trans-Martian trajectories' deficiency relative to the SLS launch vehicles have other down-stream effects as to its appropriateness for beyond-Earth orbit crewed exploration. Traditionally, crewed exploration beyond Earth orbit has focused on minimizing complexity, and therefore risk and cost, by using a heavy-lift rocket. The logic behind using a heavy-lift rocket for lunar exploration in the past was that fewer launches correlated to less risk. The Falcon Heavy's sixteen tons capability means that any lunar exploration program will have to be one of assembling pieces/parts in Earth orbit, where the Falcon Heavy's fifty three tons payload capacity can really shine. A lunar crewed mission using the Falcon Heavy would mean assembling, at necessary LEO locations, a crewed vehicle, a lander, a trans-lunar injection stage, a stage to get the crewed spacecraft and Moon lander, and possibly a separate stage to enable the crewed spacecraft to return to Earth. This means more launches and added costs and risks.

SpaceX announced its intention to send passengers on a crewed Dragon spacecraft, launched with a Falcon Heavy rocket, around the moon in 2018/2019. This would follow on the heels of the company's robotic and crewed flights to the International Space Station, and an uncrewed Falcon Heavy moon trip. NASA's own mission, which would be the first crewed flight using the SLS and its new Orion spacecraft, is scheduled for 2021. SpaceX Moon mission would bring real information on costs and acceptable risks. At the present stage, it is better to have two suppliers from a competition/cost point of view. It means that the jury is still out on the optimal solution for man missions beyond Earth orbit.

5.2 Space agencies programs

Some space agencies have adopted a "planet hopping" approach in pursuing human space exploration. What they do is to take the planetary destinations in sequence, referring to the Moon, near Earth objects, the moons of Mars, and Mars itself. There is a strong debate both within space agencies and other institutions on the optimal way forward with this approach. Some favor the Moon as the first target in a planet hopping approach. Others prefer exploration of near earth asteroids. But from the political and economic points of view, the architecture of the first move of humans into deep space would be better served by a lunar space station, leaving the search and exploitation of lunar and near earth asteroids resources to private companies.

NASA revealed a new plan for human space exploration, starting with the construction of a new space station in cislunar orbit. NASA and its international space station partners are putting the finishing touches on humanity's new home in space. If approved by governments, this future space station will effectively replace the International Space Station when it reaches retirement age in the 2020s. The most exciting idea behind this new station, destined to make humans' home orbiting the moon, is that it would provide a new foothold for future human missions to other planets in the solar system. It would also offer an initial in-space market for space resources, which may help the development of the new space industries. This future space station will effectively replace the International Space Station when it reaches retirement age in the 2020s. As the costs of building from scratch such infrastructure are so large and not be workable for any single nation, international cooperation makes sense because it will spread the large costs among several nations.

NASA engineers and their colleagues from four other space agencies have been working on the design of the new station for a few years. If everything goes as planned, the first mission will be launched in 2023, and position a robotic spacecraft called the Power and Propulsion Bus (PPB) in a stretched orbit around the Moon. Then, in the next two years, a pair of barrel-shaped modules - four-and-a-half meters wide and five meters long, and both weighing under ten tons each - will be delivered to the same orbit and bolted to the PPB. Together, they will house four astronauts up to ninety days at a time. In addition, a Russian-built airlock module will be added to the station in the mid-2020s to make it much easier for the crew to walk outside of its space home.

Along with NASA's new Orion spacecraft, the Russian next-generation spacecraft that is now under development might drop by for a visit as well. NASA sources also hint that the station will be able to accommodate an extra module, in case a "new partner" decides to come on board, most likely referring to China. In addition, the cislunar base might serve as a parking hub for the robotic (and later human) lander returning from the surface of the Moon and bringing the raw materials (oxygen and hydrogen) necessary for the life support system of the station and for fuels for spaceships returning to Earth and for those venturing beyond the Moon. The means of getting to the space station itself is still up for debate. Each crew shifts on the cislunar outpost will be accompanied by a dedicated supply ship, but NASA has not yet decided on the mode of transportation or its provider, whether public or private.

NASA plan has received many critics. Some claim that, if the goal is to build a Moon base, it should be built on the surface of the Moon. That is where the science is, that is where the shielding materials are, and that is where the resources to make propellant and other useful things are to be found. These critics say that one does not need a lunar-orbiting station to go to the Moon, nor to go near-Earth asteroids and to Mars. According to these critics, this proposal is notable for requiring a large budget to create an object with no utility whatsoever.

Although these critics are based on common sense, the question of a lunar space station is more subtle. First, the creation of a Moon base by a space agency would produce serious international frictions in space. Despite the existence of the United Nations agreement on the use of celestial bodies, it is doubtful that national governments would be bound to any considerations except to their own national interest, leading to the militarization of space. In this respect, a lunar space station built and managed by international partners, and leaving the exploitation of lunar resources to the competition of private entities is a much better solution. In this way, international frictions in space would be reduced, since it would transfer the search for raw materials to the competition among private entities and away from that of powerful and armed states.

Second, is a lunar space station better suited than the Moon to launch spaceships to other bodies in the solar system? The answer is yes, provided that the space station is located at a Lagrangian point, where Earth and the Moon gravitational influences cancel, so that spaceships need no energy to break free from Moon or Earth gravity. Third, a lunar space station can be a good testing ground to develop ways to protect astronauts from cosmic radiations[1]. Long-distance manned missions are inevitably going to have to shield astronauts from cosmic radiation. Although many materials[2] have been proposed to protect astronauts, creating an artificial magnetic field surrounding the lunar space station and, in the future, spaceships is probably the best way to protect astronauts from cosmic radiations.

This idea was first put forward in the 1960s[3], but was considered unworkable because it was reckoned that a volume of space more than hundred kilometers across would have to be shielded. Extending a field over this distance would require magnets with strengths of tens or even hundreds of Teslas, which could not be transported into space. However, according to recent research[4], these previous calculations were inaccurate because they assumed that the solar-wind plasma flows like a normal fluid. Decades of

---

[1]Cosmic radiation is mostly composed of energetic elements like carbon, oxygen or hydrogen that are present across space and have had some of their electrons stripped from them. And it is that lack of electrons which makes them radioactive, because when these elements hit something they can draw electrons away from whatever they have run into and can break molecular bonds.

[2]Originally aluminum was picked because it was relatively lightweight, and is about twenty five percent effective at blocking heavy ion radiation particles. And now people are looking into different types of polyethylene plastics and other materials that look to be about thirty three percent effective.

[3]See R.H. Levy et al. (1964); and F.W. French et al. (1968).

[4]See R. Bamford (2008)

research on nuclear fusion have instead shown that plasma is subject to all sorts of turbulent behavior not seen in normal fluids. The researchers believe it is possible to exploit this knowledge of turbulence to create a much smaller protective bubble, and have confirmed this belief, first through computer simulations and then in a laboratory experiment. These results show that a space station could, in fact, be protected using a bubble just some 100-200 m across. This, they say, would correspond to a magnet of about 1 Tesla, which would be light enough to be transported into space. When developed this technology could be applied to spaceships in long distance manned missions.

5.3 Global problems

The prospect of growing mismatch between people and resources manifested several times in human history. In the late eighteen century, population growth in Europe, the Americas, and China prompted Malthus to write his famous book Essays on Population. Malthus forecast that there would be an increasing gap between people's food demand and the capacity of land to satisfy it. The results would be increasing starvation, diseases and death. Malthus dire predictions did not materialize. In England and in Europe, three factors prevented these dire forecasts. The first one was the large migrations of people in other continents. The second one was the large improvements in farming output, and the third one was technological innovations leading to the Industrial Revolution. Hence, the greatest tests facing human society in confronting the 21st century is how to use the power of technology, and space technology is part of it, to meet the demand thrown up by the global problems facing the world in the near future. Increasing world population[5] and economic activities in developed and emerging economies are already causing substantial damages to the world ecosystem. The physical impact of population increases upon the natural environment ranges from the impact

---

[5]Population expansion is not a new phenomenon, but has acquired a tremendous acceleration in the last sixty years. In 1825, the world population totaled one billion people. In the following one hundred years, the world population doubles reaching two billion people, and between 1925 and 1976 it doubled again to four billion people. Although the increase has slowed down in recent decades because of reduced fertility rates in many countries, the world population reached 7.4 billion in 2016. While there remains a range of estimates on what the Earth population will be in 2025 and in 2050, the raw figures are daunting. Demographers expect average family size to decline in the future due to urbanization and other factors, leading to a stabilization of world population. But this will happen gradually and stabilization is expected in 2050. Because birth and death rates varies over time, demographers offer three possible estimates (high, middle, and low). Accordingly, world population can stabilize in 2050 between 10 to 15 billion.

on earth's atmosphere to agriculture, deforestation, and depletion of fossil fuels and minerals. In the medium term, extraction of critical minerals and fossil fuel resources may reach the limits of economic feasibility, and could lead to a collapse of key infrastructures unless new resources are found. Space technology, particularly mining of space resources, may come to the rescue.

5.4 The Moon's mineral resources

The Moon is the near-Earth object most extensively explored. Its physical exploration began in 1959 with the landing of Soviet Luna 2[6]. From the mid-1960s to mid-1970s, there were sixty five Moon landings, both manned and robotic. Perhaps most celebrated, the seven Apollo missions brought back to Earth 380 kilograms of samples from the lunar surface. After 1976, lunar exploration halted. It resumed in 1990 with the Japanese Hiten spacecraft. ESA sent a small, low-cost lunar orbital probe in 2003. Its primary goal was to take three-dimensional X-ray and infra-red imagery of the lunar surface. NASA sent the Lunar Reconnaissance Orbiter in 2009 to collect extensive imagery of the Moon's surface. It carried the Lunar Crater Observation and Sensing Satellite, which investigated the possible existence of water on the Moon. The People's Republic of China began a program for Moon exploration in 2007, whose long term goal is lunar mining, specifically rare earth elements and helium-3 for use as an energy source on Earth.

The Moon's exploration and analysis of the samples brought back to Earth show its geological composition to be similar to the Earth (Table 5.1).

However, there are some characteristics that cause the Moon to be quite different. The Moon lacks a significant atmosphere, which eliminates erosion due to weather; it does not have any form of plate tectonics, it has a lower gravity. The lack of living matter on the Moon causes the soil to lack organic compounds. The outermost layer of the Moon is called the crust, which extends down to a depth of fifty kilometers. This is the layer of the Moon that scientists have gathered the most information about. The crust of the Moon is composed mostly of oxygen, silicon, magnesium, iron, calcium, and aluminum. There are also trace elements like titanium, uranium, thorium, potassium and hydrogen. Base metals like iron and aluminum are bonded to atoms of oxygen and silicon, called silicates. Metals are generally found as metal oxides. Samples brought back by the Apollo missions show significant variations from sample to sample at each site.

---

[6]Luna 2 was a space probe launched by the Soviet Union and made an impact on the Moon surface in September 1959.

Table 5.1 – Abundance of Chemicals and Metals on the Moon

Components	*A 11*	*A 12*	*A 14*	*A 15*	*A 16*	*A 17*	*Luna 16*	*Luna 20*
SiO2	42.47%	46.17%	48.08%	46.20%	45.09%	39.87%	43.96%	44.95%
Al2O3	13.78%	13.71%	17.41%	10.32%	27.18%	10.97%	15.51%	23.07%
TiO2	7.67%	3.07%	1.70%	2.16%	0.56%	9.42%	3.53%	0.49%
Cr2O3	0.30%	0.35%	0.22%	0.53%	0.11%	0.46%	0.29%	0.15%
FeO	15.76%	15.41%	10.36%	19.75%	5.18%	17.53%	16.41%	7.35%
MnO	0.21%	0.22%	0.14%	0.25%	0.07%	0.24%	0.21%	0.11%
MgO	8.17%	9.91%	9.47%	11.29%	5.84%	9.62%	8.79%	9.26%
CaO	12.12%	10.55%	10.79%	9.74%	15.79%	10.62%	12.07%	14.07%
P2O5	0.12%	0.10%	0.09%	0.06%	0.06%	0.13%	0.21%	0.08%
H	51 ppm	45 ppm	80 ppm	64 ppm	56 ppm	60 ppm		
He	60 ppm	10 ppm	8 ppm	8 ppm	6 ppm	36 ppm		
C	135 ppm	104 ppm	130 ppm	95 ppm	106.5 ppm	82 ppm		
N	119 ppm	84 ppm	92 ppm	80 ppm	89 ppm	60 ppm	134 ppm	107 ppm
Rare Earth Elements					903 ppm			

Source: Beauford R. (2011); G. Kramer et al.(2003); and http://www.permanent.com/lunar-geology-minerals.html.
Note: A stand for Apollo missions. Ppm (part per million) is a way of expressing very dilute concentrations of substances. Just as per cent means out of a hundred, so parts per million or ppm means parts out of a million.

Two minerals have attracted the interest of space agencies, governments, and private companies. The first one is rare earth elements. Apollo explorers, forty years ago, brought back rocks that proved to be high in rare earth elements. The region from which these rocks originated is called the Procellarum KREEP Terrane. Concentrations of rare earth elements in the Lunar KREEP are much higher than those in Earth crust, but lower than those in North

American and Chinese mines[7]. Rare earth elements are used to build permanent magnets, metal alloys, and catalysts. If electric cars come to dominate the auto market, previously powered by the internal combustion engine, demand for these elements will skyrocket. Consequently, off world supply will become an even more interesting economic proposition. The other one is helium-3. Lunar soil[8] contained significant quantities of a remarkable form of helium Unlike Earth, which is protected by its magnetic field, the Moon has been bombarded with large quantities of Helium-3 by the solar wind. Helium-3 is a scarce element on Earth. Used in industry, it needs to be manufactured and is a product of the decay of tritium, which results from neutron bombardment of deuterium, lithium, boron, or nitrogen targets. Current industrial consumption of helium-3 is around 60,000 liters per year. The price has varied from a low of 100 dollars per liter to at around 75,000 tonnes, which could satisfy future demand for use in prospective nuclear-fusion reactors.

In the short-medium term, water and hydrogen are the most promising resources to exploit on the Moon for the space market. Water ice has been discovered on the Moon by radar back scatter at the North Pole[9] and by spectrometry at the South Pole in the Cabeus crater with an extrapolated volume for both poles of conservatively 109 metric tons. NASA scientists are developing two separate mission concepts to assess, and learn how to exploit, stores of water ice on the moon and other lunar resources. The projects, called Lunar Flashlight and the Resource Prospector Mission, are notionally targeted to blast off in 2017 and 2018, respectively, and aim to help humanity extend its footprint out into the solar system. 2,000 dollars per liter in recent years. Commercial fusion reactors using helium-3 are not yet available. Even if they become commercially viable, they will need tens of tonnes of helium-3 each year to produce a fraction of the world's power. Recoverable helium-3 resources of the Moon are estimate.

---

[7]Concentrations of rare earth elements in continental earth crust are 168 ppm, while concentrations in Mountain Pass mine and Mount Weld mines are respectively 76,000 and 130,000 ppm.

[8]A sample of soil from the rim of Camelot crater slid was brought back to Earth by the crew of Apollo 17. That sample 75501, along with samples from *Apollo 11* and other missions, were analysed in 1985 by young engineers at the University of Wisconsin who discovered that lunar soil contained significant quantities of a remarkable form of helium-3.

[9]The first major discovery was by the Chandrayaan 1–mini-SAR mission from mid-February to mid-April, 2009 was based on back scattered radar signals from the north lunar pole from mid-February to mid-April, 2009. Based on these information the volume of water discovered in the lunar north pole using radar back scatter of circular polarization ratios was estimated to be 600 million metric tons of water ice.

Lunar Flashlight aims to find water ice that would be accessible to future explorers. Lunar Flashlight is a CubeSat mission, meaning the body of the spacecraft is tiny. Lunar Flashlight would start orbiting the moon about six months after its launch, then spend another year spiraling down to get to about twenty kilometers from the lunar surface. The probe would then make about eighty passes around the moon at this low altitude, measuring and mapping deposits of water ice in permanently shadowed craters near the lunar poles. While Lunar Flashlight would eye the moon from above, the Resource Prospector Mission (RPM) plans to send a rover onto the lunar surface to get an up-close look. This rover would land at a yet-to-be-determined polar site and map surface and subsurface concentrations of hydrogen at two different locations, which would ideally be separated by at least one kilometer. RPM would use a neutron spectrometer to measure water concentrations up to one meter underground and a near-infrared spectrometer to make its surface measurements.

5.5 The asteroids' mineral resources.

Asteroids are small, airless rocky worlds revolving around the Sun, in a belt between Mars and Jupiter. They are leftovers from the formation of the solar system about 4.6 billion years ago. Early on, the birth of Jupiter prevented any planetary bodies from forming in the gap between Mars and Jupiter, causing the small objects that were there to collide with each other and fragment into the asteroids seen today. Asteroids can be as large as Ceres, which is 940 kilometers across. On the other hand, the smallest asteroid ever studied is the 2 meters space rock 2015 TC25, which was observed when it made a close flyby of Earth in October 2015. Nearly all asteroids are irregularly shaped, although a few of the largest are nearly spherical, such as Ceres.

For the purpose of this analysis, the most interesting are the near-Earth asteroids (NEA). NEA have orbits bringing them closer to Earth on a regular basis. Most NEA are likely the result of collisions between asteroids in the main asteroid belt triggered by Jupiter's gravitational influence. A few NEA may be the spent nuclei of comets. They are grouped into three categories: (i) Amors, (ii) Apollos, and (iii) Atens. Amors cross the orbit of Mars, but do not reach Earth orbit. Apollos cross Earth's orbit with a period longer than one year. Atens cross Earth's orbit with a period shorter than one year.

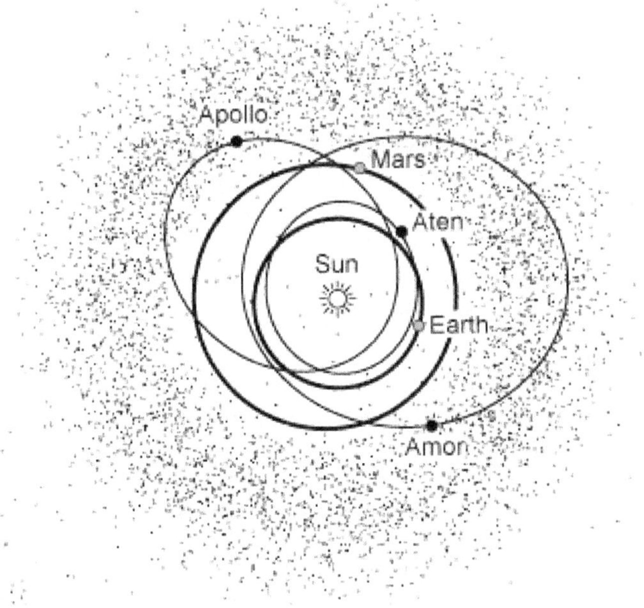

Figure 5.1 – Map of the asteroids
Credit: Professor Kenneth R. Lang, Tufts University

In the last decade, analysis of asteroids' mineral composition has advanced via: spectroscopic and dynamic studies of asteroids; analysis of meteorites fallen on Earth, and unmanned space missions to the near earth asteroids. Japan sent an unmanned mission to return a sample of material from a small near-Earth asteroid, named 25143 Itokawa, to Earth for further analysis[10]. In 2014, Japan sent another unmanned asteroid sample return mission (Hayabusa 2). The target is asteroid 162173 Ryugu. Hayabusa 2 is expected to arrive at the target in July 2018, survey the asteroid for a year and a half, depart in December 2019, and return to Earth in December 2020. The spacecraft features ion engines, upgraded guidance and navigation technology, antennas and attitude control systems. Operations at the asteroid will be similar to

[10]The mission was launched on 9 May 2003 and rendezvoused with Itokawa in mid-September 2005. After arriving at Itokawa, Hayabusa studied the asteroid's shape, spin, topography, color, composition, density, and history. In November 2005, it landed on the asteroid and collected samples in the form of tiny grains of asteroidal material, which were returned to Earth aboard the spacecraft on 13 June 2010.

those of the previous Hayabusa, but with an explosive device to dig the asteroid surface for fresh sample material[11].

The majority of known asteroids fall into three categories. Type C (carboniferous) asteroids account for around 75% of known asteroids. They are dark with an albedo between 0.03 to 0.09. These asteroids contain large amounts of carbon molecules, metals, water, and volatile. Type S (siliceous) asteroids account for around 17% of known asteroids. They are relatively bright with an albedo between 0.10 to 0.22. It is estimated that these asteroids are composed of iron-nickel mixed with silicates. Type M (metallic) asteroids account for the rest of the known asteroids. They are bright with an albedo between 0.10 to 0.18. Their composition is dominated by iron-nickel.

Ground-based astronomical observations have already made substantial progress in defining and characterizing the population of NEAs, including preliminary estimates of their mineral composition. Algorithms have been developed to define the orbital characteristics, mass of the accessible asteroids, required Δv, length of the missions, and possible launching dates of identified bodies. Results from spectroscopic and dynamic studies are confirmed by analysis of meteorites fallen on Earth. Asteroids contain valuable elements such as the platinum group metals (PGM), rare earth elements, gold and germanium. Identified minerals also include nickel, in the form of metallic nickel-iron, and cobalt (Table 5.2).

Type C asteroids contain water and volatile, which are valuable resources for in-space fuel and life-support systems. PGM and precious metals' concentrations vary with the asteroid, but are higher than in the Earth's crust. Most metals sank to the Earth's core when the planet was young, while concentrations in the asteroids are higher because of low gravity. They reflect the original distribution of gases and dust from which the solar system originated. It is estimated that asteroid's raw materials are of excellent quality. Meteorites' analysis show minerals in a free state. If confirmed, asteroids' minerals will need little machining during future mining operations, as compared to the lengthy and costly refining operations on Earth.

Asteroid exploration is planned both by space agencies and private companies. Apart from their scientific objectives, space agencies missions could be justified by the following argument. On Earth, geological knowledge of an area is usually provided by governments as a public good, and it is widely recognized as an important element for attraction of private companies and for contributing to the continued development of mineral resources. Similarly, the exploration programs of the asteroids from space agencies could open

---

[11]For a detailed technical analysis of Hayabusa 2 mission, see M. Yoshikawa (2011).

the door to exploitation of natural resources by the private sector, not only through the technologies to reach and land on the asteroids, but also through the identification of near Earth asteroids and the verification in situ of their composition, orbital characteristics, required Δv, length of the missions, and possible launching dates of identified bodies.

Table 5.2 – Abundance of PGMs and Precious Metals in Asteroids

Metals	*Asteroid LL-chondrite*	*"Good" metallic asteroid*	*"Best" metallic asteroid*	*Earth's crust*
Industrial elements				ppm
Cobalt	1.57%	0.46-0.80%	0.43-0.75%	25
Nickel	34.3%	5.6-18.0%	5,4-16.5%	120
Iron	63.7%	81.0-94.0%	82.0-94.0%	55,000
Germanium	1,020	0.06-70	0.05-35	1.8
PGM	ppm	ppm	ppm	ppm
Rhenium	1.1	1.1	2.4	0.0004
Ruthenium	22.2	20.7	45.9	0.001
Rhodium	4.2	3.9	8.6	0.0002
Palladium	17.5	12.6	12	0.0006
Osmium	15.2	14.1	31.3	0.0001
Iridium	15.0	14.0	31.0	0.0003
Platinum	30.9	28.8	63.8	0.005
Precious metals				
Gold	4.3	0.16-0.70	0.06-0.6	0.001
Rare earth elements	0.4	n.a.	n.a.	0.2

Sources: Calculated from data given by Buchwald V. F. (1975); Malvin D. J. et al. (1997); Hoashi M. et al. (1993); and R. P. Mueller et al. (2010).

Recently, NASA has approved Asteroid Redirect Mission[12] (ARM) whose goals are: (i) identification, capture, and transport of a small asteroid[13] in a stable lunar orbit; and (ii) human missions to this asteroid while in lunar orbit. The corollary goals are several: testing new propulsion systems and the difficult maneuvers of rendezvous with the asteroid; and analyzing the effects on astronauts of long sojourns in space outside Earth's magnetic field. ARM would thus enable the assessment of the mineral resources' composition of the asteroid, and the testing of methods and systems to carry out these operations. These are also valuable pieces of information for space mining companies to reduce risks of future operations on an industrial scale.

In Europe, ESA has been developing programs for the exploration of asteroids and comets and which will add to our asteroid's mineral resource database. ESA's Rosetta probe reached Comet 67P/Churyumov-Gerasimenko in 2014. The probe experienced fly-by encounters with many asteroids. These encounters were fleeting but added to our knowledge of the asteroids. ESA is sponsoring Spaceguard, a worldwide asteroid-tracking project that aims to plot the orbits of those asteroids on possible collision courses with the Earth. So far, it has logged more than 300 asteroids. Asteroid explorations are likely to feature in the Aurora Program, a joint ESA-NASA long-term program for the manned exploration of the solar system.

Private companies, such as Deep Space Industries and Planetary Resources are designing and planning exploratory missions to the asteroids. Planetary Resources is designing orbital telescopes to explore asteroids. The orbital telescopes are envisioned as the first step forward in the company's asteroid mining ambitions. Deep Space Industries is designing and planning low-cost (CubeSats) reconnaissance missions to near Earth asteroids to analyze their key aspects and bring back samples. The key aspects under inquiry are: size of the asteroid; composition; mass/density; spin; interior structure; surface structure; and accurate orbit. The final goal is to identify the best possible candidates for future mining operations.

Space agencies are likely to find synergies with private companies for asteroids' reconnaissance missions. This would liberate resources for the space agencies' strategic missions. Agreements can be reached with private space mining companies. Companies would carry out reconnaissance missions to

---

[12]NASA plans to launch a spaceship to recover an asteroid or a part of it in 2019, capturing it with an inflatable structure similar to a bag. Advanced solar-electric propulsion (SEP), called an ion thrust, will propel the spacecraft. High-efficiency ring-shaped solar panels will provide the electrical power.

[13]Size is important for security reasons. If something were to go wrong, an asteroid, whose diameter is below twenty meters, will disintegrate in Earth's atmosphere.

targeted asteroids. Private companies would share information with space agencies for a price. This would be a fraction of the costs incurred by space agencies to carry out these missions. Alternatively, private companies would exchange information for participation in manned missions, such as planned missions to visit a small asteroid brought into lunar orbit. This would enable testing mining methods and equipment, reducing the risks of future operations on an industrial scale. Both solutions are Nash equilibrium. Each player's strategy maximizes his expected utility against the other players' strategies, and no player can be better off by unilaterally changing his strategy. An important step would be to set up an adequate forum to discuss and exchange information on programs, affordability, and possible missions' time frames.

Recently, an Italian researcher[14] put forward a proposal for exploration of asteroid 3753 Cruithne[15]. It is a five-kilometer-wide near-Earth object which shares an orbit with the Earth. The Italian's plan is to send a small spaceship propelled by an electric-ion engine to rendezvous with Cruithne. Then it will deploy two smaller CubeSats to examine the asteroid more closely. Cruithne's large orbital inclination presents a particular challenge for any spacecraft to reach it, one that certainly tests the mettle of any proposed space mission. The mission goes straight from Earth to Cruithne without having to slingshot en route. This will save on transit time and complexity while offering a highly flexible launch window. The spacecraft is expected to weigh around hundred kilograms and will reach Cruithne in around 320 days. While interesting, the proposal is still a proof-of-concept and needs further elaboration to reach the stage of a real mission's content.

This proposal should be given serious consideration for two reasons. First, Cruithne is of great scientific interest. Cruithne's composition preserves its original chemical make-up -- unaltered by high internal pressures and temperatures -- telling more about how the solar system formed. Second, a survey mission will help pave the way to robotic-landing missions, and human explorations and asteroid-mining endeavors. Cruithne could contain the largest deposit of space minerals near Earth. For example, assuming an average density of 2.5 grams per cubic centimeter and around 31 ppm of platinum, the estimated platinum's mass is around 330,000 metric tonnes.

---

[14]See Pergola P. (2013).

[15]Duncan Waldron discovered Cruithne in October 10, 1986. Cruithne has a normal elliptic orbit around the Sun, with periodic revolution around the Sun almost equal to that of Earth. Because of this, Cruithne and Earth appear to "follow" each other in their paths around the Sun. This makes Cruithne a co-orbital object, whose orbit, viewed from Earth, is described as a horseshoe. Its inclination to the ecliptic is 19.82^0. Cruithne has an albedo of 0.12, which makes it an S type asteroid. Gravity at its surface is estimated at 1.622 meters per square seconds.

5.6 Concepts of mining and processing

The technologies needed to carry out the mining process in space are being developed by the new space industries. Since these technologies have proprietary natures, it is difficult to access these data. But possible methods have been suggested in the literature. Table 5.3 gives possible products and process of water and volatile, and minerals founded on the Moon and different types of near earth asteroids.

On the Moon, water and volatile, nickel, rare earth elements, and helium-3 are the most promising materials. Several different rovers and mining drills have been tested in Hawaiian field tests, and demonstrated how future missions could excavate water from lunar regolith or underground ice pockets. Several water processing technologies have been studied. Water on the Moon poles and in craters is present in the chemically unbound form of ice and requires only modest heating (to 100° C) for vaporization and collection. Later, one can work up to the more difficult things. One promising technology takes advantage of the chemistry of the moon dirt - or regolith - by adding hydrogen, which then reacts with iron oxide in the moon dirt to produce water. Such hydrogen reduction reactors heat the regolith to about 1,000 degrees Celsius so that the proper chemical reactions can occur. A process known as electrolysis can then split the extracted water into pure hydrogen and oxygen, either for rocket fuel or astronaut air supplies.

Table 5.3 – Products and processes

Moon	*Materials*	*Process*
	H_2O, CO_2	Thermal dehydration
	NiFe metal	Carbonyl process
	Rare earth elements	Thermal and chemical
	Helium-3	Thermal, mechanical
Asteroid type		
Carbonaceous	H_2O, CO_2 NiFe metal	Thermal dehydration Magnetic separation
Ordinary chondrite	NiFe, silicates	Magnetic separation
Metallic	NiFe Platinum Group Metals	Magnetic separation carbonyl process

Credit: author.

Nickel can be processed by a Mond process. This process involves the fact that carbon monoxide combines with nickel readily and reversibly to give nickel carbonyl. The pure nickel reacts with carbon monoxide at 50–60 °C to form the gas nickel carbonyl, leaving the impurities, including iron and cobalt, as solids. No other element forms a carbonyl compound under the mild conditions used in the process. The mixture of nickel carbonyl and Syngas is heated to 220–250 °C, resulting in decomposition back to nickel and carbon monoxide: Nickel can then be removed from the gas again at higher temperatures, perhaps in an attached printer.

Rare earth elements (REE) could be processed through a series of steps. First, material is passed through a cascade of physical processing steps including crushing, screening, grinding and flotation finally producing a pre-concentrate from which products with different rare earth oxide (REO) contents are derived in subsequent thermal and chemical processing steps. The REE mineral concentrate is digested in hydrochloric acid and the resulting REE laden liquor is passed through a sequence of solvent extraction with several mixer-settler steps. Final products of this element-selective extraction processes are either separated REO or other products. Although considerable lunar soil would have to be processed to extract helium-3, the mining costs would not be high by terrestrial standards. Automated machines might perform the work. Extracting the isotope would not be particularly difficult. Heating and agitation release gases trapped in the soil. As the vapors are cooled to absolute zero, the various gases present sequentially separate out of the mix. In the final step, special membranes would separate helium-3 from ordinary helium.

J. S. Kargel[16] has analysed extraction and processing methods for asteroids' minerals. On some types of asteroids, material may be scraped off the surface using a scoop or auger, or for larger pieces, an active grab. A mine can be dug into the asteroid, and the material extracted through the shaft. This requires precise knowledge to engineer accuracy of astro-location under the surface regolith and a transportation system to carry the desired ore to the processing facility. These operations are much less massive than on Earth, as they do not need heavy machinery for extraction and ore's transport to the processing center. Asteroid minerals need less machining since metals are already in a free state. The ore is ground and sieved into different sizes, and transported to the processing center.

Asteroids such as carbonaceous chondrites that contain hydrated minerals, water and other volatile, which can be extracted simply by heating. A water extraction test in 2016 by Honeybee Robotics used asteroid regolith simulant developed by Deep Space Industries and the University of Central Florida to match the bulk mineralogy of a particular carbonaceous meteorite. The simu-

---

[16]See Kargel J. S. (1994).

lant was physically dry (i.e., it contained no water molecules adsorbed in the matrix of the rocky material), heating to about 510 °C released hydroxyl, which came out as substantial amounts of water vapor from the molecular structure of phyllosilicate clay and sulfur compounds. The vapor was condensed into liquid water filling the collection containers, demonstrating the feasibility of mining water from certain classes of physically dry asteroids.

Ordinary chondrites are rich in iron and nickel, which can be extracted by magnetic separation. After grinding, the streams of material are put through an "impact grinder" or "centrifugal grinder" whereby a very rapidly spinning wheel accelerates the material down its spokes and flings it against an impact block. Silicate impurities attached to the free metal are shattered off. A centrifugal grinder may be used after mechanical grinding and sieving, and before further magnetic separation. Magnetic fields then separate the nickel-iron metal granules from the silicate grains. Repeated cycling through the magnetic field gives highly pure bags of free nickel iron metal.

The platinum group metals, present in metal asteroids, could be extracted by the Mond process, also known as the carbonyl process. This process involves combining carbon monoxide with nickel readily and reversibly to give nickel carbonyl. The pure nickel reacts with carbon monoxide at 50–60 °C to form the gas nickel carbonyl, leaving the impurities, including iron and cobalt, as solids. No other element forms a carbonyl compound under the mild conditions used in the process. The mixture of nickel carbonyl and Syngas is heated to 220–250 °C, resulting in decomposition back to nickel and carbon monoxide: Nickel can then be removed from the gas again at higher temperatures, perhaps in an attached printer, leaving platinum group metals and gold as a residue. Finally, ion exchange techniques, distillation, and solvent extraction will separate the platinum group metals from gold.

5.7 The economics of space mining

Space mining is like any other mining investment project, more uncertain for sure, but conceptually similar. A profit-maximizing firm will invest to maximize the surplus of expected revenues over expected costs. Each stream is discounted to present value. The net present value of this project is given by the following equation:

$$NPV = [\sum (p_t s_t - oc_t q_t)e^{-\delta t}] - C \quad (1)$$

where:

p_t = price of the mineral

q_t = mass produced

oc_t = unit operational costs

s_t = annual sales

δ = discount rate

C = investment costs, including R&D, exploration costs, and launching costs

t = time of the project

The volume of the mineral ($\sum s_t$) brought back to the reference market is derived from the Tsiolkovsky rocket equation, assuming that the propellant for the return trip is produced on the celestial body. The equation relates the Δv with the effective exhaust velocity and the initial and final masses of a rocket. As the space vehicle's mass is small compared to the mass of the mineral plus propellant, with a certain approximation, the last one is equal:

$$\sum s_t = M_{produced} \times e^{-\Delta v/ve} \qquad (2)$$

where $M_{produced}$ is the mass produced on the celestial body, Δv is the change in velocity required for the return to the reference market, and ve is the exhaust velocity. With some transformation (see Appendix to this chapter), equation (1) can be written as:

$$NPV = (\Sigma\,(p_{ts}\, x\, ((M_{mpe} \times f \times t_{stm}) \times e^{-\Delta v/ve})/n) \times e^{-\delta ts}) - (oc_{tm} \times (M_{mpe} \times f \times t_{ttm})) \times e^{-\delta tm}) - (C\, x\, e^{-\delta t}) \quad (3)$$

where t_s is the time interval between the arrival of the minerals on Earth and the end of sales, n is the number of years of sales, tm is the time of stay on the celestial body plus the time from LEO to the celestial body and return to LEO, and t is total time from the beginning of the project to the end of sales. M_{mpe} is the mass of the equipment, f is the throughput factor expressed as the ratio of kilograms per day of the materials treated per kilograms of equipment, and t_{stm} is the time of stay on the celestial body.

5.7.1 In-Space as a reference market

The Moon has abundant mineral resources. These are relevant to a hierarchy of future applications, beginning with the use of lunar materials to facilitate human activities on the Moon itself, and progressing to the use of lunar resources to underpin a future industrial capability within the Earth-Moon system. The reference market would be the new international station and fuel for satellites in Earth orbit. Water and volatile can be used to supply oxygen for life support to the new international station and propellant for spaceships returning to Earth and the ones venturing in deep space beyond the Moon. marketable with the in-space market. Iron, nickel, titanium and other minerals can be used to build consumable by using 3D printing.

A potential market exists also at low earth and geostationary orbits. Located about 22,000 miles above Earth, GEO is home to more than 400 satellites, many of which deliver such essential services as weather reports, cell phone communications, television broadcasts, government communications and air traffic management. LEO is the orbital highway for many satellites providing Earth observation and weather tracking. Many satellites are healthy and in good operating condition in-orbit, and are able to operate beyond their fifteen years design life. For these satellites, the limiting lifetime factor is the remaining fuel on board to maintain attitude control and orbital position. Supplying these satellites with fuel would extend the life of these satellites.

The main competitive advantage of the Moon materials stems from the comparative requisite changes in velocity, as the main competitor for the space mining company is water, volatile, and consumables brought directly from Earth. Getting the fuels and other consumables from the Moon into the new space station and satellites in low earth and geostationary orbits will be relatively cheap. Because of the peculiarities of celestial mechanics, such a haul requires just 1/14th to 1/20th of the fuel it takes to bring these materials up from Earth. This is clearly shown in Table 5.4. It shows that the changes in velocity for the different transfers from Earth to LEO, from LEO to the new international space station, and from the Moon surface to the new international space station and from there to Earth orbit. Thus energy needed to transport water, volatile, and consumables from Earth is higher than the one needed to transport these materials from the Moon to the in-space market.

Table 5.4 – Δv Required for Lunar Transfer

Transfers	*Δv (km/s)*
Earth surface to LEO	8.0
LEO to escape velocity	3.2
LEO to the new international space station	6.3
Moon surface to the new international space station	2.3
International space station to LEO (with aerobraking)	2.4

Source: Lewis, J. S. (1996).

At present, no feasibility study exists on the construction of a Moon base and in situ resource utilization. Potential revenues of the Moon base stem from resource mining (oxygen, rocket fuel), energy (helium 3 mining for nuclear fusion), and space tourism. Solar power would keep operating costs down. Minimizing the initial investment (C in equation (1)) is a prerequisite to a successful Moon project. Initial investments relate to the costs of building a lunar base and the development of a transportation vehicle from the Moon base to the new international space station. Sophisticated robots could prepare the landing site prior to the arrival of astronauts. These initial investment costs could be reduced by the lowering of transportation costs from Earth. Competition among different launcher producers is already working to achieve this goal.

Entirely new classes of space vehicles would be needed, ones that operate only at and beyond Earth orbit, such as vehicles for orbital transfer and satellite repair. Today launch systems must be designed to withstand the punishing effects of high-speed atmospheric drag, pressure, vibration, and heating that occur on the way to space. Protecting the rocket and its payload adds enormously to launch costs. But a vehicle that is designed from the start to operate only in space is not bound by the same design rules. Hence the lower costs of these vehicles.

The investment costs could be reduced by employing inflatable structures for the Moon base. Constructed of multi layer fabrics shielded with Kevlar or other strong materials and banded by steel exoskeletons, these structures could provide most of man habitation, storage, and transportation requirements. They would be both lighter and less expensive than traditional spacecraft. A number of companies have done extensive R&D on such inflatable space structures, including Boeing and Bigelow Aerospace, which has even lifted two test modules to low Earth orbit. These structures would be covered by regolith from the Moon to protect his inhabitants from cosmic radiations.

Cooperation of private companies with space agencies may reduce the investment costs. Since the Moon has many mysteries to unravel that are important for understanding the origins of Earth and the entire solar system, scientific knowledge is another argument lowering the costs of a permanent lunar base. The lunar surface is also an ideal platform for astronomical observations. It is stable, with low gravity, and an absence of wind or other significant forces; so it's an ideal place for a telescope. The Moon base could provide the habitat for scientists operating these instruments. Hence the investment costs of establishing a lunar base could be shared with governments that provide funding for scientific research.

5.7.2 Earth as a reference market

As previously outlined, over the next twenty five years extraction of critical minerals and fossil fuel resources may reach the limits of economic feasibility. Space mining of minerals, particularly PGM and rare earth elements could come to the rescue of the industrial civilization. PGM and rare earth elements can be found in near earth asteroids and the Moon. This competitive advantage of asteroids' minerals stems from the comparative requisite changes in velocity (table 5.5). Velocity changes from low-Earth orbit (LEO) to the Moon are higher than the ones to near-Earth asteroids (NEA). Similarly, velocity changes from the lunar surface to LEO are higher than the ones from NEA to LEO. Thus less energy is needed for asteroid missions than for Moon missions, and hence the competitive advantage of NEA for minerals' supply to the Earth market.

Table 5.5 – Δv Required for Transfer of Minerals to Earth Orbit

Transfers	*Δv (km/s)*
Earth surface to LEO	8.0
LEO to escape velocity	11.2
LEO to the Moon landing	6.3
LEO to NEA	4.0
Moon surface to LEO (with aerobraking)	2.4
NEA to Earth transfer	1.1

Source: Lewis, J. S. (1996).

There are two ways to mine the near earth asteroids. A first one is to return a small asteroid to Earth or Moon orbit, where the minerals will be extracted and sent to Earth. The Keck Institute for Space Studies[17] (KISS) has conducted a feasibility study for retrieval of a small asteroid. According to the KISS study, the cost for a future mission to identify and return a five hundred tons asteroid to low earth orbit is about 2.6 billion dollars, ignoring the costs to develop the infrastructure necessary to process the materials in the asteroid. This study states that asteroid mining is feasible as long as three major advances are in place: an efficient solar/electric propulsion system, a campaign to discover and target potential asteroids, and the establishment of a human presence in lunar orbit.

Planetary Resources estimates that a single thirty meter long platinum-rich asteroid could contain 25 to 50 billion dollars worth of platinum at today's prices. They estimate that, once the proper infrastructure is in place, there is potential for significant profit. These results are based on very simplifying assumptions and exclude the costs of transportation of metals from lunar or low earth orbits to earth. They should be treated with caution, but they offer important indications on the project feasibility. A major concern with this approach is the risk posed by the asteroid to Earth. It is estimated that, if

[17]See Keck Institute for Space Studies (2012)

something goes wrong, an asteroid below twenty meter long would disintegrate in entering Earth atmosphere. Bigger asteroids would not, posing substantial risks to Earth. Careful considerations should be given by the authorities to asteroid mass and the stability of its orbit around the Moon before granting license to these projects.

The second way of asteroid mining in their natural orbits was first discussed D. W. Cox[18]. Later on, D. L. Kuck and M. J. Sonter[19] analysed it in depth. To bring the mass of metals to the Earth market, a firm sustains a stream of costs over several time periods. Initially, the mining company incurs costs related to: (i) identifying the targeted asteroid; (ii) and research and development of mining and refining equipment for their use in space. As in the case of the Moon, space agencies exploration of asteroids would reduce these costs for private space mining companies.

Larger investment costs occur before launching the mission, and depend on the costs of constructing the spaceship and the necessary equipment, power supply, cargo container, and instrumentation and control:

$$C = (M_{mpe} + M_{ps} + M_{i\&c}) * (p_e + p_l) \quad (4)$$

where C is the cost of investment, and M_{mpe}, M_{ps}, and $M_{i\&c}$ are respectively the mass of the equipment (M_{mpe}), power supply and cargo container (M_{ps}), and instrumentation and control ($M_{i\&c}$), and pe and pl are costs of construction and launching the spaceship. These costs depend on the expected mass to be returned to Earth orbit, and the asteroid orbital characteristics. The mass returned to Earth orbit depends on the length of the mining season and on throughput factor. The length of the mining season varies and depends on the orbital characteristic of the asteroid. Aten asteroids have a quasi-circular orbit, hence longer mining seasons, while Apollo asteroids have elliptical orbits and shorter mining seasons. Mining season can also vary, depending on the eccentricity and orbit inclination of the asteroid. Low eccentricity targets can have longer mining seasons, while high eccentricity targets must have short mining seasons. Throughput factor is the ratio of mass of regolith or rocks treated per day to the equipment mass. It depends on engineering choices concerning mining and processing equipment and the mining season's length. Longer mining seasons imply less demanding specifications on mining and processing equipment and lower costs.

Apart from determining the mining season's length, the asteroid's orbital characteristics influence the spaceship's costs. Asteroids with low eccentricity and low inclination may need transfer with continuous thrust. Asteroids with

[18]See Cox, D. W. et al. (1964)

[19]See Kuck D. L. (1995); and Sonter, M. J. (1997).

high eccentricity and those with low eccentricity but high inclination require impulsive transfer. In the first case, the electric propulsion, such as the ion propulsion, is an option. In the second case, nuclear propulsion is the only choice. A recent study[20] suggests that the development of this technology could cost between 2.5 to 3 billion dollars over five years. Research and development of an electric propulsion engine are well advanced, and the next generation of ion engines will offer advanced capabilities with lower development costs compared to nuclear technology.

The above analysis indicates that, from the costs point of view, the optimal choice is an Aten asteroid with low eccentricity and inclination. The outlays are bell shaped, with the peak rate occurring at the time of the spaceship's construction and launch. After launching and until return to Earth orbit, spending is consumed by operational costs, which are minor compared to investment costs. A final cost occurs when transferring metals from Earth orbit to Earth surface. But no unique pattern of expenditures exists. The expenditures' pattern depends on management decisions on the rate of spending during research and development.

Management can spend at higher rates in the early period to bring the project's operations to an earlier date. But, when time is saved by increasing the rate of spending beyond a level sustaining economies of scale, total development costs increase. Research and development is a heuristic process and often involves significant uncertainties. Alternative technical approaches with finite success probabilities can be explored in series until success emerges, but this takes time. Overall expected time can be reduced by running technical approaches concurrently. But this increases the expected project costs because more approaches are run, some of which ultimately prove unnecessary. The relationship between expected research and development time and expected total development costs is thus convex to the time and costs coordinates. As time is compressed, total development costs rise at an increasing rate.

A profit maximizing firm will adjust time allocated to the R & D program and asteroid's choice to maximize the surplus of discounted expected revenues over total expected costs. Expected revenues depend on how other companies in the market react to the presence of a new firm. The PGM and rare earth markets are oligopolistic ones. There are few PGM and rare earth elements producing regions worldwide and few mining companies. Opportunities exist for new entrants due to expected high demand and supply constraints forecast to prevail in the future. But the new entrant ought to devise a sale strate-

---

[20]See Howe S. et al. (2012).

gy to maximize its profit in time. No unambiguous solution exists. The outcome will depend on specific assumptions about other companies' reactions. The economic literature has analysed these problems, and it suggests that an equilibrium solution exists if the new entrant's annual supply is small and information perfect. This solution is often criticized as unrealistic because it is rare that firms select quantities simultaneously with no ability to know their competitors' choices.

Heinrich von Stackelberg pointed out that results will be different if firms chose quantities sequentially. Stackelberg games are classic examples of the bi-level optimization problems often encountered in game theory and economics. These are complex problems with hierarchical structure where one optimization task is nested within the other. In a multi-period strategy with imperfect information, a sequential equilibrium prevails. Each company infers information on other companies' behavior by their past actions and adjusts next period's offers. The adjustment will be incomplete in any single period, but over the longer run a final state of balance results. From a practical point of view, the off-world miner could reach agreements with a service company, such as Johnson Matthey in London, which would store the metals returned to Earth. The service company would then sell the metals through a series of forward contracts, thus acting as liquidity provider and sales agent.

5.8 Conclusions

This chapter provided an overview of the rationale for and the feasibility of the Moon and near earth asteroid mining based on current technology and information. It concluded that the mining of the Moon and of the near earth asteroids could shift from science fiction dreams to world-changing reality a lot faster than one thinks, as there is one thing that seems to repeat throughout history: a tendency to overpredict what will happen in the next year, but to vastly underpredict what will happen in the next ten to twenty years. A conjunction of events makes it likely that space mining would become a reality in the next twenty to twenty five years. These events are the continuous lowering of the costs of accessing space, pressures from population on Earth mineral resources and the environment, space agencies programs, and the awakening of individuals and private industries, who see potential profits in the expansion of the economic sphere beyond Earth orbit.

If NASA proposal to build a space station in lunar orbit, in cooperation with other space agencies, is carried out in the next ten years, it is likely that Moon mining for water, hydrogen, and some metals for the in space market would be the first venture in space mining. A feasibility study for the new space station does not yet exist. But one can take as a starting reference the costs of building of the International Space Station, whose overall costs amounted to around 150 billion dollars. Over a ten year period, this means about fifteen billion dollars per year. Although this amount is most likely be on the lower side, the costs of building the new space station could be within the budget capabilities of space agencies participating in the program without major increases in their budgets. Mining the Moon would serve the space station needs for oxygen for life support system, propellant for spaceships on their return to Earth, and consumables. Recently, the United States government has officially approved the plan for a robotic lunar landing of California-based Moon Express, which aims to fly commercial missions to Earth's nearest neighbor and help exploit its resources.

The exploration of the near earth asteroids by space agencies and private companies could be completed over the next five to ten years. From a technical point of view, the capabilities to explore the near earth asteroids are there. The financing of asteroid exploration by space agencies would not require major increases in their actual budgets, if synergies with private companies are found. Asteroid exploration would then open the way to their mining by the private sector, which could most likely occur in the early 2030s. The financing of asteroid mining by private companies could be achieved through financial and equity markets, provided that an international legal framework on the exploitation of space resources is in place, and that some of the cost issues raised in the previous paragraphs are resolved.

The economic and political benefits of space mining would be large. It would avoid the effects of the growing mismatch between world population and mineral resources, and increase the resources available for economic growth on Earth. The establishment of an international legal and institutional framework for a peaceful evolution of the space economy beyond Earth orbit would make different space faring nations feel secure that their own companies could participate in these ventures, and that all countries will be able to directly reap the benefits of space resources. A more stable world resulting from space mining would then generate the resources to proceed with the next step of planet hopping, i.e. the exploration and colonization of Mars, which will be the subject of the next chapter.

Appendix - Technical Note

This appendix focuses on the determinants of the net present value:

$$NPV = [\sum (s_t \times q_t - oc_t \times q_t) \times e^{-\delta t}] - C \quad (1)$$

and particularly on the volume of the metals brought back from the asteroid to LEO ($\sum s_t$). As the space vehicle's mass is small compared to the mass of the metals and assuming that the propellant for the return trip is produced on the asteroid, with a certain approximation, the last one is equal:

$$\sum s_t = M_{produced} \times e^{-\Delta v/ve} \quad (2)$$

where $M_{produced}$ is the mass produced on the asteroid, Δv is the change in velocity required for the return by orbital transfer, and v_e is the exhaust velocity.

The mass produced on the asteroid is equal to:

$$M_{produced} = M_{mpe} \times f \times t_{stm} \quad (3)$$

where M_{mpe} is the mass of the equipment, f is the throughput factor expressed as the ratio of kg per day of the materials treated per kg of equipment, and t_{stm} is the time of stay on the asteroid (mining season). Substituting (3) into (2) gives:

$$\sum s_t = (M_{mpe} \times f \times t_{stm}) \times e^{-\Delta v/ve} \quad (4)$$

This equation shows the mass of the metals returned to LEO. It depends on technical factors, the length of the season mining, the propulsion system adopted, and the change in velocity necessary for the transfer between the two orbits.

It remains to find Δv, the change in velocity needed for the return trip. Δv is a scalar quantity dependent on the desired trajectory and not on the mass of the space vehicle. It is calculated as the sum of the Δv's required for the propulsive maneuvers during the mission. It determines how much propellant is required for a vehicle of given mass and propulsion system. The Δv_r to insert into the return trajectory is small, due to the low gravity of NEA. But Δv_{ic} to correct for inclination change must be added.

The asteroid in the hypothetical case is an Aten type of asteroid. This mission profile assumes a target body with a semi axis major less than 1; a quasi circular orbit (an Aten); and a mining season commencing at perihelion and running until aphelion. This mission assumes Hohmann transfer to rendezvous with the target asteroid at its perihelion and with a near-aphelion departure after half an orbit stay time. The Hohmann transfer orbit is an elliptical orbit used to transfer objects between two circular orbits of different radii in the same plane.

An example of the Hohmann transfer is illustrated by the following figure 5.2.

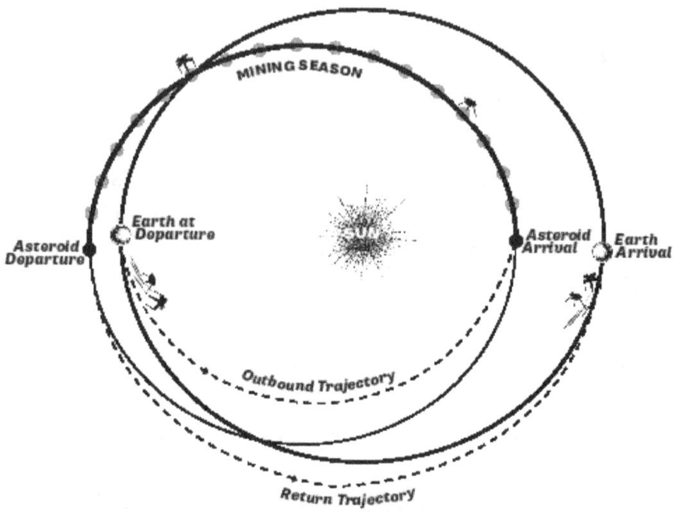

Figure 5.2 – Mission to an Aten asteroid
Credit: Author

Following Hohmann[21], the minimum change in velocity for the return trip is equal to the difference between the required velocity to enter the transfer orbit (v_1) and the transfer ellipse velocity (v_2):

$$\Delta v_r = (v_1 - v_2) \quad (5)$$

where:

$$v_1 = \sqrt{}((\mu / ((1.5 \times 1011) \times Q) \times (1 - e_{transfer})) \quad (6)$$

and

$$v_2 = \sqrt{}((\mu / ((1.5 \text{ x } 1011) \times Q)) \times (1 - e_{tr}) \quad (7)$$

where:

$\mu = G \times m_s$

G =Newton gravitational constant

---

[21]See Thomson (1986).

m_s = mass of the sun

Q = the aphelion of the asteroid in AU

$e_{transfer} = (1 - q)/(1+q)$

q = the perihelion of the asteroid in AU

e_{tr} = the eccentricity of the asteroid.

Using the orbital characteristics of the asteroid 2004 MN4, v_1 is equal to 37.62 km/s. v_2 is equal to 36.9 km/s. Thus the change in velocity for transfer orbit insertion is 0.73 km/s. However, we have to add the velocity change required for the inclination change. It depends on where the inclination change occurs, and the heliocentric velocity at that point. This velocity change can be approximated by:

$$\Delta v_{ic} = (0.52 \times \text{inclination}) \quad (8)$$

which, in the hypothetical case, is equal to 1.73 km/s. The total velocity change for the return to Earth transfer is then equal to:

$$\Delta v_T = \Delta v_r + \Delta v_{ic} \quad (9)$$

or 2.46 km/s.

We assume that the fuel needed for the return trip is extracted from the asteroid. Let us assume use of a nuclear thermal rocket. Nuclear rockets were being actively developed under the NERVA project. Conceptually, it works by pumping hydrogen through the core of a nuclear reactor to heat it to produce exhaust as a reaction fluid. A full-scale operating nuclear rocket was tested and achieved a specific impulse (Isp) of 850 seconds. It nearly attained flight readiness before funding was withdrawn. We assume that the nuclear thermal rocket achieves Isp of 850 seconds. Since exhaust velocity is equal to the specific impulse time the standard, not local, acceleration of gravity, the exhaust velocity of around 8.3 km/s. If one assumes that the mass of the spaceship as negligible with respect to the mass of the payload, from the rocket equation, the following holds:

$$M_{start} - M_{finish} \times e^{\Delta v/ve} \quad (10)$$

In our hypothetical case, M_{finish} is about 800 tonnes and $e^{\Delta v/ve}$ is 1.36. The amount of fuel needed for the return trip is about 320 tonnes. The total time of the mission is equal to the time needed for research and development and the construction of the equipment and the spaceship (t_1), plus the transfer time from LEO to the asteroid and return to LEO. We assume that t_1 is about 1.1 years, and the time from LEO to the asteroid and return to LEO is approximately equal to 3T/2, where T is the orbital period of the asteroid. In our hypothetical case, this is equal to 1.3 years and the total time of the mission is 2.4 years.

The equation for the net present value of the project can then be rewritten as:

$$NPV = (\Sigma\ (\ p_{ts}\ x\ ((M_{mpe} \times f \times t_{stm})\ x \times e^{-\Delta v/ve})/n)\ x\ e^{-\delta ts})\ -\ (oc_{tm}\ x\ (M_{mpe} \times f \times t_{ttm})) \times e^{-\delta tm}) - (C \times e^{-\delta t}) \qquad (11)$$

where t_s is the time interval between the arrival of the minerals on Earth and the end of sales, n is the number of years of sales, tm is the time of stay on the asteroid plus the time from LEO to the asteroid and return to LEO, and t is total time from the beginning of the project to the end of sales.

Equation (11) shows that the total project time is crucial to the feasibility of the project. The longer is the project time, the lower the net present value of the project. This equation indicates that an Aten asteroid is most likely the best target for asteroid mining. Other types of asteroid, such as Apollo and Amor, need longer project time. An example of a Hohmann transfer for an Apollo asteroid is given by figure 5.3.

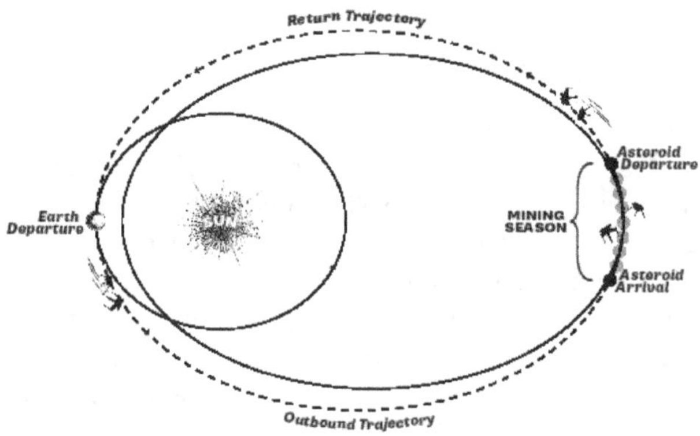

Figure 5.3 – Mission to an Apollo type of Asteroid
Credit: author

These missions imply Hohmann transfers with rendezvous near but before aphelion for least changes in velocity; a short aphelion-centered mining season; and a post-aphelion departure for Earth-return, for least Δv_r. Such missions have longer project duration with respect to Aten's missions. The use of Hohmann ellipse transfer and short mining season imply that mission duration approximates the orbital period of the asteroid. Transfer time is thus normally longer for Apollo asteroids than Aten ones. Longer project duration affects negatively the net present value. Shorter mining seasons need tougher demand specifications for the equipment for the same mass of metals returned to Earth orbit increasing investment costs.

Bibliography

Bamford R. et al.: *"The interaction of a flowing plasma with a dipole magnetic field: measurements and modeling of a diamagnetic cavity relevant to spacecraft protection"*, Plasma Physics and Controlled Fusion, Volume 50, Number 12, 2008

Beauford R.: *"Mining the Moon for Rare earth Elements"*, in http://www.rareearthelements.us/ree_in_space, 2011.

Buchwald V. F.: *"Handbook of Iron Meteorites: Their History, Distribution, Composition, and Structure"*. University of California Press, Berkeley, 1975.

Cox D. W. et al.: *"Islands in Space: The Challenge of the Planetoids"*, Chilton Books, Philadelphia, 1964.

French F.W. et al.: *"Plasma Radiation Shield: Concept and Applications to Space Vehicles"*, Journal of Spacecraft 5:5, pp. 570-577, 1968.

Hoashi M. et al.: *"Palladium, Platinum and Ruthenium in iron meteorites and their taxonomic significance"*, Chemical Geology 106 1993 pp. 207-218, 1993.

Howe S. et al.: *"Ground Testing a Nuclear Thermal Rocket: Design of a Subscale Demonstration Experiment"*, American Institute of Aeronautics and Astronautics, AIAA 2012-3743, 30 July 2012.

Kargel J. S.: *"Metalliferous asteroids as potential sources of precious metals"*, Journal of Geophysical Research, 99 25 No E10 pp. 21129-21141, Oct 1994.

Kramer G. et al.: *" The composition of Kreep: A detailed Study of Kreep Basalt 15386"*, Lunar and Planetary Science, vol. XXXIV, 2003.

Kuck D. L.: *"The Exploitation of Space Oases"*, Princeton Conference on Space Manufacturing, Space Studies Institute, 1995.

Levy R.H. et al.: *"Plasma Radiation Shielding"*, AIAA Journal 2:10, pp. 1835-1838, 1964.

Lewis, J. S.: *"Mining the Sky"*, Helix Books, New York, 1996.

Malthus T. R.: *" An Essay on the Principle of Population as It Affects the Future Improvement of Society"*, ed. J. Bonar, New York, 1965.

Malvin D. J. et al.: *"Chemical classification of iron meteorites. Multi-element studies of 43 irons, resolution of group IIIE from IIIAB, and evaluation of Cu as a taxonomic parameter"*, Geochimica et Cosmochimica Acta 48:785–804.

Mankins J. C.: *"SPS-ALPHA: The First Practical Solar Power Satellite via Arbitrarily Large Phased Array "*, 2011-2012 NASA NIAC Phase 1 Project, Final Report, September 2012.

Mueller R. P. et al.: *"Regolith as a Resource in Solar System Human and Robotic Exploration"* paper presented at the 61st International Astronautical Congress, Prague: International Astronautical Federation, 2010.

Pergola P.: *"Small Satellite Survey Mission to the Second Earth Moon"*, Advances in Space Research, vol. 52, issue 9, pp. 1622-1633, November 2013.

Sonter M. J.: *"The technical and economic feasibility of mining the near-earth asteroids"*, Acta Astronautica 41(4-10), 637–647, 1997.

Thomson, W. T.: *"Introduction to Space Dynamics"*, Dover, NY, 1986.

Yoshikawa M.: "Asteroid Exploration Mission Hayabusa 2", paper (in Japanese) presented to the 11th Symposium on Space Science, February 2011.

Chapter 6

The development of the space economy beyond Earth orbit: Mars

"The rocket will free man from his remaining chains, the chains of gravity which still tie him to this planet. It will open to him the gates of heaven."

Wernher von Braun

Once space infrastructure and space mining of the Moon and/or the near earth asteroids are in place, moving human space exploration and the space economy beyond the Moon and near earth asteroids is the next challenge, and Mars is the next target. Since the Apollo program ended, human exploration of Mars has been a goal of space agencies and other organizations over the intervening decades. In fact, these organizations have put forward mission proposals[1] since the 1950s, but none of them has gone beyond the study stage. This long period has had, however, some benefits since it allowed ideas on the architecture of the mission to mature; misconceptions to be overcome; and research on necessary technical innovations to take place.

But a fundamental flaw of these studies was that, while technical details and the architecture of the mission were extensively considered, less explored were the political, economic, and financial motivations behind the establishment of a permanent human base on Mars. Sending exploratory missions to Mars to take a few pictures, planting a flag, and returning a few samples of

[1]Mission proposals include among others: (i) Project Empire, a study commissioned by the Marshal Space Flight Center and awarded to General Dynamics and Lockheed (1959); (ii) von Braun's Mars Project (1969); (iii) The Martian Expeditionary Complex by the Soviet Union (1969); (iv) NASA Space Exploration Initiative (1989); (v) Mars Direct by R. Zubrin and D. A. Baker (early 1990s); (vi) NASA Design Reference Mission (early 1990s and 2000); (vii) Mars Piloted Orbital Mission by the Russian Space Agency (2000/2005); (viii) Aurora Program by ESA (2001), ESA-Russian Plan (2002); and Vision for Space Exploration by the United States (2004).

rocks to Earth would encounter many resistances and ultimately fail. As Buzz Aldrin once stated[2] that:

> "When we set out to land people on the surface of Mars, I think we should as a nation, as a world, commit ourselves to supporting a growing settlement and colonization there. To visit a few times and then withdraw would be an unforgivable waste of resources."

This chapter looks first to the technical and economical feasibility of establishing a permanent human base on Mars, and, second, it analyses different human exploratory mission proposals in preparation for the establishment of a permanent human base on Mars.

6.1 The construction of a permanent base on Mars and its sustainability.

The key questions on building a permanent base on Mars are his technical and economic feasibility, and what benefits Earth will derive from this venture.

6.1.1 Mars physical characteristics

Although Mars Mars is the most Earth-like planet in the solar system, his characteristics are very different from Earth. Mars is less dense than Earth, having about fifteen per cent of Earth's volume and eleven per cent of Earth's mass, resulting in about thirty eight per cent of Earth's surface gravity. Like Earth, Mars has differentiated into a dense metallic core overlaid by less dense materials. Current models of its interior imply a core with a radius of about 1,794 kilometers, consisting primarily of iron and nickel with about sixteen to seventeen per cent sulfur. The core is surrounded by a silicate mantle and crust that formed many of the tectonic and volcanic features on the planet, which appear to be dormant. Based on data sources from robotic missions, scientists think that the most abundant chemical elements in the Martian crust, besides silicon and oxygen, are iron, magnesium, aluminum, calcium, and potassium. The elements titanium, chromium, manganese, sulfur, phosphorus, sodium, and chlorine are less abundant, but are still important components of many minerals in rocks and of secondary minerals (weathering products) in the dust and soils (the regolith).

The geography of planet Mars is complex. The main features are shield volcanoes, lava plains (mostly in the northern hemisphere) and highlands with a large number of impact craters and deep canyons. Martian rocks seem mostly basalt, although a portion of the Martian surface seems richer in silica than

[2]See interview, *Scholastic*, Nov. 17, 1998.

typical basalt. The plains are similar to rocky deserts on Earth, covered by red sand, with rocks and boulders scattered all around. The soil is essentially regolith, rich in finely-grained iron oxide dust, which is thinner than on the Moon. Its granulometry and composition is more variable from place to place, due to water erosion in ancient times when water was flowing on the surface, and to wind erosion. The soil of Mars is basic. This, together with the results of the old Viking probes seem to exclude the possibility of finding life-forms on the surface of the planet.

Table 6.1 summarizes the main physical characteristics of the Earth and Mars.

Table 6.1 – Main characteristics of Mars and Earth

	Earth	Mars
Mass (10^{24} kg)	5.9736	0.64185
Volume (10^{10} km^3)	108.321	16.318
Equatorial radius (km)	6378.1	3396.2
Polar radius (km)	6356.8	3376.2
Ellipticity	0.00335	0.00648
Mean density (kg/m^3)	5515	3933
Surface gravity (m/s^2)	9.81	3.71
Surface acceleration (m/s^2)	9.78	3.69
Escape velocity (km/s)	11.19	5.03
Solar irradiance (W/m^2)	1367.6	589.2
Orbital semiaxis major (10^6 km)	149.60	227.92
Perihelium (10^6 km)	147.09	206.62
Aphelium (10^6 km)	152.10	249.23
Mean orbital velocity (km/s)	29.78	24.13
Orbit inclination (deg)	0.000	1.850
Orbit eccentricity	0.0167	0.0935
Length of day (hours)	24.0000	24.6597
Axis' rotation inclination (deg)	23.45	25.19

Credit: author

Mars offers limited protection from cosmic rays due to the absence of a planetary magnetic field. Mars' current magnetic field is very weak, with strengths of at most about 1500 nanotesla. Earth's magnetic field, by comparison, varies up to around 65000 nanotesla, or more than 40 times stronger than Mars'. Earth's magnetic field is supported by an internal dynamo. Mars must once have had a dynamo, which would have magnetized its rocks, but then the dynamo shut down. Observations show that parts of the planet's crust have been magnetized, suggesting that alternating polarity reversals of its dipole field have occurred in the past. Analysis of the Mars Global Surveyor[3]), suggests that plate tectonic activity occurred on Mars four billion years ago, before the planetary dynamo ceased to function and the planet's magnetic field faded.

Mars has a thin atmosphere[4] and low average surface temperatures (between -140 ° C and 20 ° C). Carbon dioxide is the main component of the atmosphere (95.3%). Nitrogen and oxygen are around three per cent. Methane has been detected in the Martian atmosphere with a concentration of about 30 ppb (part per billion). It occurs in extended plumes, and the profiles imply that the methane was released from discrete regions[5]. The atmosphere is quite dusty, containing particulates about 1.5 μm in diameter which give the Martian sky a tawny color when seen from the surface. The lengths of the Martian seasons are about twice those of Earth's because Mars' greater distance from the Sun leads to the Martian year being about two Earth years long. Martian surface temperatures vary from lows of about −143 °C at the winter polar caps to highs of up to 35 °C in equatorial summer. The wide range in temperatures is due to the thin atmosphere which cannot store much solar heat, the low atmospheric pressure, and the low thermal inertia of Martian soil.

---

[3]Mars Global Surveyor is orbiting Mars and mapping the planet's magnetism (or lack thereof) with its magnetometer. Every once in a while, anomalies are found, where some magnetized substance is buried beneath the surface. These anomalies are thought to have come up from a once-magnetized core and kept their magnetism when the planet lost its overall magnetism.

[4]The pressure at the ground is less than one hundredth of the atmospheric pressure on Earth and has much variability with altitude and latitude. It varies from around 0.3 millibar on Olympus Mons to over 11.6 millibar in the depths of Hellas Planitia. It has a mean surface level pressure of 6.36 millibar, ranging from 4.0 to 8.7 millibar depending on the season.

[5]In northern midsummer, the principal plume contained 19,000 metric tons of methane, with an estimated source strength of 0.6 kilograms per second. The profiles suggest that there may be two regions of local methane source. The first centers near 30°N 260°W and the second near 0°N 310°W. It is estimated that Mars must produce 270 tonnes per year of methane. See Mumma M. J.; et al. (February 20, 2009).

The comparatively large eccentricity of the Martian orbit has a significant effect. Mars is near perihelion when it is summer in the southern hemisphere and winter in the north, and near aphelion when it is winter in the southern hemisphere and summer in the north. As a result, the seasons in the southern hemisphere are more extreme and the seasons in the northern are milder than would otherwise be the case. The summer temperatures in the south can be up to 30 °C warmer than the equivalent summer temperatures in the north. Mars has the largest dust storms in the Solar System. Dust storms tend to start in well-known places in the southern hemisphere as it warms up. These can vary from a storm over a small area, to gigantic storms that cover the entire planet. They tend to occur when Mars is closest to the Sun, and have been shown to increase the global temperature.

Liquid water cannot exist on the surface of Mars due to low atmospheric pressure, which is less than one per cent that of Earth's. The present-day inventory of water on Mars is estimated from spacecraft imagery, remote sensing techniques (spectroscopic measurements, radar), and surface investigations from lander and rovers. Water exists as ice, with a small quantity present in the atmosphere as vapor. Water ice is visible at the surface at the north polar ice cap. Abundant frozen water may be present beneath the permanent carbon dioxide ice cap at the south pole. Large quantities of water are thought to be trapped underground. In the northern hemisphere an ice permafrost mantle stretches down from the pole to latitudes of about 60° and large quantities of water ice have been observed both at the poles and at mid-latitudes. It may also be present in the shallow subsurface at more temperate latitudes, and in the interiors of caverns and lava tubes, which may also provide protection from micrometeorites, cosmic radiation, solar flares and high energy particles.

6.1.2 Technical feasibility

Several organizations are studying a permanent base[6] on Mars. The Mars Foundation is funding studies for the architecture's design; of technologies related to the life-support systems, and for equipment to build the base. Significant challenges for Mars habitats are shielding from intense solar radiation and maintaining an artificial environment. Long term permanent habitats require large volume (i.e.:greenhouse) and thick shielding to minimize the annual dose of radiation received. This type of habitat is too large and

[6]Principally the Mars Foundation. Other organizations, such as NASA, are simulating technologies and equipment needed to build the Mars base. Simulations are conducted in zones with temperatures close to those on Mars, such as the Antarctic and northern Canada.

heavy to be sent to Mars, and must be constructed making use of some local resource. One of these proposals is the Adobe[7] infrastructure made with Martian regolith. Martian regolith is a free resource available in unlimited quantities on the surface.

The Adobe exterior serves as a radiation shield, and a thermal insulation layer. It provides complete protection against dust storms and micrometeoric impacts. The Adobe exterior houses an inflatable membrane that functions as a self-contained pressure space for the colons to inhabit. Its construction requires on the surface a regolith drill/pump, a brick baker, plastic super-adobe bags to hold soil, an inflatable half-cylinder membrane, and a plexiglass sheet supply. Other possibilities include excavating subterranean spaces, or sealing the ends of an existing lava tube[8]. Most studies concentrate on an underground permanent base or caves. In particular, large lava tubes have been detected on the planet which can be quite useful for installing habitats even of large size. An underground base and caves give good protection against cosmic and solar radiation, wind, and dust.

Besides radiation concerns, another issue with humans living on Mars is gravity. Humans have evolved in a 1g gravity environment. Once on Mars, surface gravity is only thirty eight per cent of that on Earth. The relatively low Mars surface gravity should not pose the same threat of orthostatic intolerance or bone fracture as ISS astronauts experience on return to Earth. But we simply do not know the effects of long term or permanent stay on Mars on the human body. In order to thrive on the Red Planet in the future, it is vital that thorough research into the Martian environment and its interaction with the complex human body is carried out now. In particular, long-term isolation studies that simulate not only the journey to Mars, as in the case of the Mars 500 experiment, but also daily life on the surface of Mars as a human settler are needed. Medical experiments investigating the environmental effects of the Martian environment, such as reduced gravity, should also be carried out.

Some have suggested that an underground gravity wheel should be central to any long term Mars base plans so that Earth-like 1g gravity is created. Fig-

[7]Adobe is the Spanish word for mud brick, made from natural building materials such as sand, clay, water, and fibrous or organic material.
[8]See http://www.zaarchitects.com/en/other/103-mars-colonization.html (2012)

ure 6.1 shows how the outer wall of the wheel[9] (the floor where people walk) must be tilted to give an effective 1g gravity on a person standing inside the wheel. The centrifugal force caused by the gravity wheel rotation causes an outward force on a person that is horizontal to the Mars surface. But the Mars native gravitational force pulls vertically downward on a person. So a person standing in the gravity wheel has both forces acting on him or her simultaneously. To deal with this, the floor of the gravity wheel is tilted until these combined forces pull on a person exactly perpendicular to the floor.

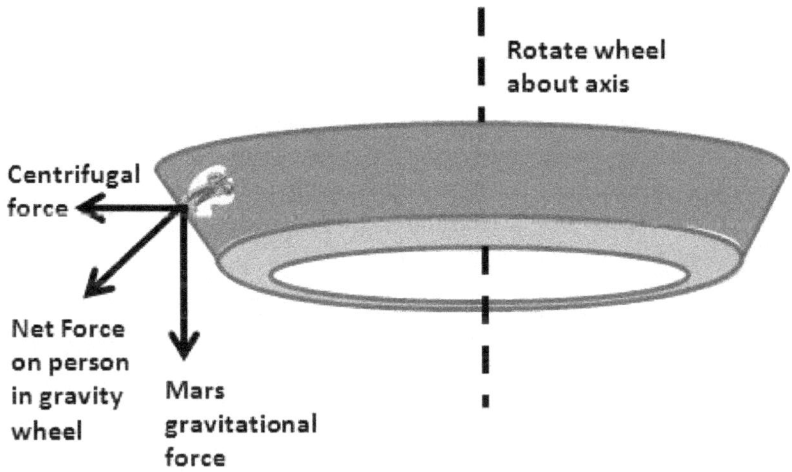

Figure 6.1 – Rotating wheel
Credit: author

An important component of the base's architecture will be the greenhouses where agricultural commodities are produced. Plants need air, soil, water, and nutrients, all of which exist to some extent on Mars. The challenges are in making the elemental needs of plant life available to plants. The first and most obvious problem to introduce human-sustaining agriculture to Mars is the nature of Martian regolith. Early studies of Martian soil showed it contained life-sustaining elements such as iron, potassium, magnesium, and sodium. Scientists as late as 2008 proclaimed Martian soil to be workable for plant

---

[9]The idea of a rotating wheel to emulate Earth's gravity has been around since the start of the 20th century. Wernher von Braun envisioned one version. The wheel must be changed so that pods where people live are tilted downward. This enables the wheel to add more gravity to existing Mars gravity. The downward angle of the pods and the wheel speed can be adjusted to give 1g net gravity.

growth. But that declaration was premature. Later, more thorough analysis of Martian soil derived from wider geographic areas leads to the opposite conclusion. Martian regolith is now thought to be saturated with caustic superoxides such as potassium hydroxide. Ways to remove superoxides from the Martian regolith are under study.

Another challenge is related to the lack of nitrogen and carbon dioxide in the Martian atmosphere. Overcoming this problem is not impossible. Carbon dioxide is part of the Mars regolith, so releasing it from its bonds is an only technical issue. The regolith can then be used to produce enough gas for plants' growth and to generate significant atmospheric pressure in greenhouses. A third problem is related to the lower gravity. While research on the International Space Station suggests plants can grow in micro gravity, scientists do not know how reduced Mars gravity might affect different Earth crops. Further research at ISS is needed. Finally, any pressurized underground greenhouse will block most of the sunlight reaching plants, so supplemental light is needed. These are not insurmountable problems, but need solution before sending out the construction team.

The use of in-situ resources will be a key driving factor to sustain human presence on Mars. There are four main strategies to produce resources for life support system and propellant: (1) bring CH_4 and O_2 to the surface of Mars; (2) bring H_2 and exploit Martian atmospheric CO_2 to produce CH_4 and O_2; (3) bring CH_4 and exploit Martian atmospheric CO_2 to produce O_2; and (4) exploit Martian atmospheric CO_2 and Martian water ice to produce CH_4 and O_2. There is no consensus on the preferred option, although the forth option would be the least costly. Whichever strategy is followed, the main components of an in-situ resource utilization plant to produce oxygen are: a reactor which reduces CO_2 by means of hydrogen; a water separator to remove the water from the CO; and an electrolyser able to separate oxygen and hydrogen from water. Initially, all these equipment have to be brought from Earth.

Robust power systems are essential for the base survival, mainly for heating and life support devices, but also for transmissions, computers, scientific instruments, and propellant production. A permanent base with a population of hundred or more will likely require in the range of several megawatts. Several sources of energy are available on Mars, from solar to nuclear and geothermal energy[10]. For solar power, Mars has an average solar radiation of 589 W/m^2 much lower than the one on Earth. Many think that solar power is not

[10]Mars has been volcanically active in the recent geological past, it is highly probable that hot underground hydro thermal reservoirs exist on the Red Planet. Once such reservoirs are found, they can be used to supply the settlers with abundant supplies of both water and geothermal power.

sufficient and propose nuclear power, although this point is controversial[11]. The true alternatives are solar, geothermal, and nuclear fission power systems, or a combination of them to compensate when solar power is not working. Mars has its geothermal power resources, which offer potential for large numbers of locally created electricity generating stations in the 10 MW (10,000 kilowatts) class. Nuclear fission is the other alternative. The nuclear power system shall include the following main subsystems: reactor; thermal-electrical conversion system; cold source (radiator); and power conditioning and distribution systems. These elements cannot be designed, transported and installed independently because they are closely bound and, initially, should be transported from Earth.

A system should be put in place to transport cargo and personnel from spaceships to Mars surface and vice versa. As the descent to Mars surface is one of the most complex problems of Mars' missions and could be very expensive, an alternative way is a space elevator. A space elevator consists of a cable with one end attached to the planetary surface and the other end in space. Stronger gravity at the lower end and outward/upward centrifugal force at the upper end result in a stationary cable being held up under tension. Space elevators reduce costs since loading and unloading to and from spaceships in orbit are mechanical operations. Their implementation on Earth is impossible with present technologies, because we have not yet the materials required to build a cable with enough compression strength to support its own weight. Materials for an Earth elevator are at least one order of strength too weak at this time. Mars gravity is around 0.378 of Earth, so materials are still too weak. But Mars has some interesting options allowing the construction of a space elevator.

---

[11]With the help of energy specialists from the Massachusetts Institute of Technology, NASA commissioned a study of how future manned Mars settlements can be powered. The MIT researchers assessed thirteen different energy generation systems, and compared solar and nuclear options. Interestingly, if positioned in the correct location, solar arrays might function just as well, if not better, than the nuclear options. It would appear that a large solar panel array can match nuclear generators, only if they are situated at a latitude of 0-40° north of the Martian equator. Southern latitudes have much less solar energy available for most of the year. A Mars mission should be able to transport several two meter-wide rolls of thin-film solar panel arrays. Rolling out an array of these thin-film rolls could supply ample energy to a colony. Solar array's old foe (dust) should not be too much of a problem as dust storms tend to start in well-known places in the southern hemisphere as it warms up, so it should not be too difficult to avoid them.

As suggested by several researchers[12], Mars' unique configuration and its nearby moon Phobos allows the construction of a simple system of space elevators. The system is composed of one space elevator binding Phobos to Mars while a second extends from Phobos outwards into space. The first elevator ends in the upper Mars atmosphere. Because Phobos is around 6,000 kilometers above Mars surface and well below geostationary orbit, the tether is within reach of today's advanced materials. The second elevator extends above Phobos so that a spaceship can dock at its end. Small craft will leave the Mars' surface to conjoin with the first elevator. They will need a much smaller Δv than getting to orbit. The first space elevator will then transport the materials to Phobos. From Phobos they are sent onward to the spaceship docked at the end of the second space elevator. Personnel and materials brought from the spaceship will follow the same route in reverse. This approach reduces the costs of transporting materials and personnel from the Mars' surface to spaceships and the other way around.

6.1.3 The sustainability of the Mars base.

Technical feasibility indicates that, while Mars is not a suitable holiday place, it is not impossible for humans to live there. It is true that most of the required technologies are already here. But, pushing them to the limits imposed by the unforgiving requirements of that endeavor, and having to combine them into the most complex space system ever conceived is a challenge that requires answers to two questions. The questions to be answered are about the hows and whys humanity should move to Mars.

The hows

Freeman Dyson is one of the few surviving giants among visionaries from the space race of the 1960s. In his book[13] "*Disturbing the Universe*", he presents heretical ideas, such as space colonization and exploitation of space natural resources. He examines two scenarios. The first one explores the feasibility of a habitat around the Sun, which uses the Sun's energy to sustain life inside the habitat. Public money finances this project. The second scenario analyses the activities of small groups of settlers, who use the asteroids' resources to support their operations. His preferences lie with the latter scenario as shown in his own words:

> "I did some historical research on the costs of the Mayflower journey, and of the migration of Mormons in Utah, and I think that we can go into space with a much lower cost. A cost of about 40,000 US$ per per-

---

[12]See L. M. Weistein (2003).
[13]See Dyson F. J. (1979).

son (in 1978 dollars or 143,254 in 2013 dollars) would be the goal to be achieved; in terms of real wages, it would make it comparable to the colonization of America. Unless the cost is brought down to that level, these projects are not very interesting to me, because otherwise it would be a luxury that only governments could afford."

Two general approaches can be easily extended to Mars colonization. The first approach involves extensive international cooperation and planning by governments. Going alone by any single nation entails prohibitively high costs. Furthermore, we know from the United Nations Outer Space Treaty that no nation can claim sovereignty over celestial bodies, as it would create the conditions for a militarization of space as powerful nations would compete in space on the logic of their own national interests. International cooperation would indubitably lower the costs for any single nation participating in the cooperation. However, those who advocate such international cooperation and planning have little conception of the difficulties that their proposals create.

Today, almost every space policy refers to international cooperation if for no other reason than it is the means to undertake space activities that otherwise could not be afforded. But changing economic and political situations during space projects that often need over a decade of commitment, the lack of transparency or trust, in particular the intentions of some countries in relation to the peaceful use of outer space, are policy challenges worth noting. It is true that a genuine international cooperation was realized around the International Space Station. But that involved only two major superpowers. Moreover, it was realized in a period in which a profound crisis hit Russia, which had all the interest in keeping his space program alive through international cooperation.

The international cooperation by governments on a Mars permanent human base would involve a larger group of countries, including China and India. The present scale of geopolitical terrestrial conflicts among nations is such that does not bode well for international cooperation. If this situation persists in the future, these impediments would pose serious obstacles to a genuine international cooperation on Mars. Furthermore, it is plausible that major space faring nations will give priority to their citizens in participating to the Mars base. If anything is evident it is that major space faring nations would abide to the rules established by themselves in the international agreement, but the order of preference outside the agreement would be to serve their own needs. This would limit the participation of smaller states and hence their eventual benefits. No real resistance from smaller states would be possible, creating the conditions for the emergence of second class nations.

The second approach is more promising. The scale of international cooperation by governments will be more limited, focusing on human Mars exploratory missions in preparation for the building of the base and the arrival of the settlers. Crucial elements of this second approach are the costs of building the base and of the voyage for each single settler. The cost of building the base could be mitigated if the astronauts of the initial exploratory missions would leave the equipment and the infrastructure of their temporary base camp for the initial settlers to survive while enlarging the base camp. This would constitute a once and for all subsidy by the international community to the future settlers.

The cost of voyage for each individual settler is also a crucial component of this mission. Recently, Elon Musk has outlined his ambition to send humans to Mars as a "backup drive" for civilization[14]. Although Musk paper is full of vision but short in details, he presents a crucial component of the architecture of the mission: the lower costs of traveling. His target is to bring that cost to 200,000 dollars, which is very close to Freeman Dyson's estimates. Musk suggested some ideas at the 2017 International Astronautical Congress on how these costs might be achieved. The answer had a lot to do with simplifying his company line up, and rendering redundant some of its current products. To do that, he would replace all of its current launch vehicles, including Falcon 9, and the larger Falcon Heavy, along with its Dragon transportation capsule with one ship and one booster. His plan is to achieve large cost savings by concentrating efforts on one flexible craft, which will also have 100 percent re-usability for more cost advantages. He predicts that the fully reusable system of the new vehicle, when complete, will actually cost less than any current launch system in operation on a per launch basis, while having far more capability. This could help SpaceX to make money with the vehicle using the kinds of launches it is operating today, in preparation for Mars missions, while dramatically cutting down on redevelopment costs. It is yet not known whether the targeted cost could be achieved, but it is a promising way to further explore.

If cost of traveling is substantially reduced, individual settlers can finance by themselves their voyages and personal belonging, including a space suit to walk on Mars. It remains the question of financing the enlargement of the temporary base camp left by preparatory exploratory missions. Investment costs would include equipment for life system and transportation on the ground, power supply, in-situ resource utilization, and a space elevator. It is impossible to give now a precise figure for investment costs, but the base's

[14]Paper presented to 67th International Astronautical Congress in Guadalajara, Mexico, 2016.

legal status will help to finance these investments. The settlers should seek dominion[15] status for the base under the aegis of the United Nations. The base should be granted autonomous governmental powers for internal management. A United Nations treaty would then define relationships with Earth, particularly for trade and security. Trade with Earth would be enforceable by having the United Nations penalize any import from Mars made anywhere on Earth in defiance to the trade agreement. The trade agreement's enforceability makes property claims on Mars secure.

The investments would then be financed by a mix of public and private funding. Given this venture's risk profile, only international financial institutions can issue debt instruments guaranteed by the Martian government. Revenues generated by the base will repay these long-term loans. The Martian government will then sell property rights on Martian soil to individual investors to generate initial revenue. The only land that exists on Mars is open, and there is an immense amount of it. It might seem worthless because it cannot be exploited right away, but land can be bought and sold if private property rights exist. For example, the Martian government grants mining rights to private groups to survey land. Such claims are tradeable on the basis of their future speculative worth.

The whys

Two questions are explored here: how economically sustainable would be the base, and what are the benefits occurring to Earth. Initially, the Mars base's population can be estimated between fifty to hundred persons. In time, one can expect the population to grow, but, given the Mars environment, it is dubious that it can grow in the millions, at least in the foreseeable future. Initially, the required skills will be mainly technical, geological, bio/geochemical, agronomical, and medical. Other skills, such as administrative, organizational, legal and logistics, should be assigned to individuals on the basis of their individual cross training.

The base will not have for a long time a division of labor to make it self-sufficient. It will need to import from Earth manufactured goods and raw materials, including fertilizers or traditional manure to produce agricultural commodities, since Mars soil is poor in nitrogen and other components necessary for agricultural production. These supplies' mass is not large and their

---

[15]In English common law, dominions were autonomous political systems nominally under the British Crown. They had specific forms of fully responsible government (not to be confused with the nature of representative government). These institutions were able to legislate on every matter, except in the fields of foreign affairs, defense, and international trade. These powers remaining under the control of the United Kingdom's Parliament.

prices on the Earth are not exorbitant. At present, the transport costs are high, but in our scenario they would be substantially reduced. Mars current account balance can be brought into equilibrium by: exporting precious raw materials from Mars' moons and other degradable materials; servicing spaceships in terms of fuel and maintenance; and innovations.

The two moons of Mars are Phobos and Deimos. While many Martian probes provided images and other data about Phobos and Deimos, only few were dedicated to these satellites[16]. Astronomical observations and flybys indicate that Phobos and Deimos both have much in common with carbonaceous C-type asteroids, with spectra, albedo, and density very similar to those of C- or D-type asteroids. If this is confirmed in the future, the moons of Mars could supply hydrated minerals, water and volatile, which could be a source of oxygen, precious minerals, and propellant. Precious minerals could be traded with Earth, if profitable. Water could be a source of oxygen and hydrogen. Oxygen could be used for life support systems on the base and for spaceships traveling between Earth and Mars, and those traveling to the outer solar system. Hydrogen and volatile could be supplied to the same spaceships. While fuel costs are great if sent from Earth, Martian hydrogen offers a competitive supply.

Another resource that abounds on Mars is deuterium, a heavy isotope of hydrogen. Deuterium could be an exportable commodity for sale and use on Earth. Contrary to common thinking, deuterium is an expensive raw material on Earth with a value of around 10,000 dollars per kilogram which is fifty times more precious than silver. Once nuclear-fusion reactors become commercial on Earth, its price will increase and will make deuterium a marketable product. Agricultural products are valuable commodities to exchange with spaceships traveling between Earth and Mars, and beyond Mars. When space mining has been extended to the asteroid belt, miners will need a steady

---

[16]Two probes under the Soviet Phobos program were successfully launched in 1988, but neither conducted the intended jumping landings on Phobos and Deimos due rocket failures. In 1997 and 1998, the Aladdin mission was selected as a finalist in the NASA Discovery Program. The plan was to visit both Phobos and Deimos, and launch projectiles at the satellites to collect ejects. These samples would be returned to Earth for study three years later. Ultimately, NASA rejected this proposal. In 2007, the European Space Agency and EADS Astrium proposed and developed a mission to Phobos in 2016 with a lander and sample return, but this mission was never flown. Since 2007 the Canadian Space Agency has been considering the prime mission to Phobos with orbiter and lander. Since 2008 NASA Glenn Research Center began studying a Phobos and Deimos sample return mission. Since 2013 NASA developed the Phobos Surveyor mission concept with an orbiter and a small rover that is proposed to launch sometime after 2023.

supply of food. It must be sent either from Earth or from Mars. The latter has a decisive competitive advantage.

Spaceships' maintenance is another marketable resource. Mars has minerals necessary to produce components of spaceships and contain ores that would be very useful to settlers. On Mars, vast volcanic landscapes are called "Large Igneous Provinces" (LIP). Such places are sources of nickel, copper, titanium, iron, platinum, palladium, and chromium. Mars' Tharsis region, which contains a group of giant volcanoes, is considered to be a LIP. Besides heat generated by molten rock, Mars has had much heat produced when asteroids impacted its surface making giant craters. The area around a large impact may take hundreds of thousands of years to cool. During that time, ice in the ground melted dissolving minerals, then depositing them in cracks or faults that were produced with the impact. Studies on the Earth have documented that cracks are produced by these impacts and that secondary minerals veins are filled in the cracks. Images from satellites orbiting Mars have detected cracks near impact craters. Areas of aqueous and low-grade thermal alteration have been found by the Opportunity Rover on the rim of Endeavor crater. So, the area around martian craters may contain a variety of minerals that were produced as an indirect result of an impact.

Another product marketable on Earth is innovation. It is known from history that a frontier society stimulates innovation for its own survival[17]. A fair system of property rights and representative governments were also key factors for innovation and economic development in frontier societies. On Mars, labor shortages, conjoined with a richly technological culture, would drive the ingenuity of the settlers to produce waves of inventions. The settlers' individualistic culture and their technological culture will help in creating institutions ensuring technological innovation. These inventions, patented on Earth, could contribute to finance trade imbalance with Earth.

Innovations are also the main product behind the why a permanent human base would be profitable to Earth. These innovations will predominately be in energy production, agriculture, automation, robotics, biotechnology, and medicine. Nuclear fusion research would be a priority for Mars settlers, as Mars has large domestic resources of deuterium fuel for fusion reactors. Deuterium is five times more common on Mars than it is on Earth, and tens of thousands of times more common on Mars than on the Moon. As water is a scare resource on Mars, development of technologies to reduce water consumption and water retention would be a priority for Mars settlers. Creating smart plastics to enable control of the spectral quality and balance of light in

[17]Labor shortages prevailing in the colonial era of 19th-century North America spurred "Yankee ingenuity" to produce waves of innovations.

the greenhouse for the next generation of protected cropping technology would be also a priority. Although we have gained some insight on the effects of micro gravity on the human body, thus far we have no knowledge of the effects of living in the presence of low gravity, such as on Mars. Long term presence on Mars would allow to gather necessary information and develop measures to adjust to these conditions, including advancement in medicine for cardiology, bone diseases, cancer, and other diseases related to living on Mars.

All these innovations and others would revolutionize and raise Earth living standards as 19th century American inventiveness changed Europe and the rest of the world. Ray Allen Billington, a leading contemporary historian of America's frontier experience, succinctly describes the American desire for mobility by touching on aspects of the nomadic culture familiar to virtually all Americans[18]. The great frontier began with the discovery of America and spread until Europeans had colonized every uninhabited or sparsely inhabited land within their reach: the Americas, Australia and New Zealand, Southern Africa, and Siberia, to mention only the largest. The American frontier experience was unique, as was each of the others. According to Billington, the uniqueness of each was produced by the particular pattern of interaction of three factors: (1) the environment upon which the frontier process operated; (2) the character of the peoples who were involved in the frontier experience, and (3) the institutions created by the unfolding experience.

These three factors would be present in the Mars frontier experience. We have already commented on the first two factors. Hereafter we will focus on the third one. If the base is granted autonomous governmental powers for internal management, the settlers' individualistic culture will help to set up a fair system of property rights, competition laws, and representative government necessary for the evolving rhythm of Mars life as each generation has had to respond to different challenges confronting it: initial settlement and growing number of bases according to population growth. In this society, creation of new opportunities on many levels of society, including new occupations to be filled by new immigrants who have the skills to do so regardless of nationality, race, gender, family background, social class, and personal influence would contribute to the maintenance or extension of equality in the social order of the Mars society.

Are all the above, technical feasibility and sustainability, justifying the decision of creating a permanent human base on Mars? A tentative answer is yes. But the jury is still out there. A final decision will depend on the results of a

[18]See R. A. Billington (1966).

human Mars exploratory mission to verify in situ the technical feasibility of human settlement on Mars.

6.2 Human exploration of Mars

Establishing a permanent human base on Mars will need careful planning, starting with an initial human exploratory mission. The main goal of this mission is to: (i) explore the feasibility of human settlement on Mars; and (ii) gather and analyze data for the base's location. Scientific goals are not excluded, but they are secondary to the main goal. The exploratory mission's goals help in defining the mission architecture's main parameters. The main mission's architecture would be the following. At least four robotic spaceships must precede the human mission to Mars. They will follow a low-energy transfer trajectory to Mars. These spaceships will contain fuel for the return trip, a descent/ascent vehicle, pre-assembled materials necessary for astronauts' survival, and the tools for exploration.

Once these spaceships are in Mars' orbit, the human mission composed of at least six crew members will depart for Mars on a spaceship. Once all the spaceships are in Mars orbit, a subset of astronauts will descend to the surface. Their stay time on Mars depends on the mission's goals and should be long enough to fulfill them. The others will stay on board the spaceship to carry out scientific experiments and survey Mars and its moons from orbit. They will also manage the communications between Mars and Earth.

This section will draw heavily on different proposals of a human mission to Mars. It explores five key elements of this mission: propulsion, trajectory, shielding from cosmic radiation during interplanetary travel, descent/ascent vehicle, landing site, and equipment necessary to fulfill the mission goals.

6.2.1 Propulsion

Rockets are momentum machines. They spew gas out of nozzles at high velocity causing the nozzles and the rocket attached to them to move in the opposite direction. So rockets are governed by Tsiolkovsky's rocket equation:

$$\Delta v = v_e \ln (m_0/m_1) \qquad (1)$$

where v_e is exhaust velocity, m_0 is mass including the propellant, and m_1 is mass without the propellant. The rocket equation has three variables. They can be cast as energies: (i) energy expenditure against gravity (Δv or change in rocket velocity); (ii) energy available in the rocket propellant (exhaust velocity or specific impulse); and (iii) propellant mass fraction (how much propellant compared to the rocket mass). Given any two of these variables, the third one becomes fixed.

Many proposals of a Mars exploratory missions considered two propulsion systems for a manned spacecraft: chemical and nuclear. Chemical rockets produce thrust by expelling the exhaust at high speed with the exhaust usually being a gas created by high-pressure combustion of the propellants, hydrogen and oxygen. It passes through nozzles that use the gas' heat energy to accelerate the propellant. The exhaust velocity generated by a chemical engine is limited because the fuel provides both heat and thrust. Nuclear propulsion comprises both thermal and electrical. In nuclear thermal propulsion, the reactor heats a propellant (usually hydrogen) that is allowed to expand through a nozzle. In nuclear electric propulsion, the reactor drives a turbine to produce electricity, which is used to create a plasma which is accelerated through the nozzle. A nuclear rocket maintains separation between the heat source and fuel, allowing for greater exhaust velocity with less propellant usage.

Both chemical and nuclear thermal propulsion has been studied with great details by a group of NASA specialists to define the NASA reference architecture for a manned Mars mission[19]. There are several advantages of nuclear engines over chemical engines. First, they can achieve higher velocities and reduce travel time. These studies estimated that, with nuclear engines, total travel time could be reduced to three to six months with respect to eighteen months with chemical engines. This limits exposure's risks to cosmic radiation for astronauts and the spaceship's sensitive components. The drawback is that the spaceship's insertion into Mars orbit requires complicated maneuvers[20] (deceleration). But this choice becomes necessary if radiation during time spent in space proves to be a show-stopper. Second, a nuclear based spaceship, on its return from Mars, will not re-enter Earth's atmosphere from a hyperbolic trajectory. On arrival, it will stay in LEO or in higher orbits where it can be refurbished and reconditioned for the next Mars missions. Multiple journeys can thus be performed in a more cost-effective way. This is particularly important if the main long-term goal is to build a permanent base on Mars.

The nuclear thermal propulsion should thus be a leading propulsion system option for human Mars missions. Main challenges associated with the use of nuclear-thermal propulsion are the following: (i) radiation; a nuclear engine gives off large amounts of radiation. Before considering these engines, effective shielding mechanisms must be developed. The goal is to limit crew's

---

[19]The first NASA report has been released in 1997 and has been followed by several studies and addendum (NASA 1997, Borowski (2009), DRA5 2009, DRA5 2014).

[20]The spaceship needs a specific velocity to orbit Mars. This velocity is less than the one needed to continue to orbit the Sun in the transfer orbit. Hence the spaceship has to decelerate to be captured by Mars gravity.

exposure to a dose of ten REM. By comparison, the limit for civilians is a dose of 150 REM, and for military personnel 500 REM. The bad news is that this radiation shielding adds to the spaceship mass; (ii) testing a nuclear-thermal propulsion device is like testing a nuclear-power plant with the primary circuit open. Today, nuclear systems must have at least three confinement barriers: fuel cladding; primary circuit envelope; and a confinement building for the reactor. Nuclear-thermal propulsion has a single confinement barrier: fuel cladding separating the uranium and fission products from the environment. An environmentally safe nuclear-thermal-propulsion system is thus a complex problem. Complexity could constitute a barrier to these engines' development if not handled properly.

A second type of nuclear propulsion is nuclear-electric propulsion. Nuclear-electric propulsion is composed of two main parts: a fission-based power-generation unit; and an electric propulsion module. A variety of thrusters use the electricity generated by the nuclear reactor. Ion thrusters use electric fields to accelerate ions to high velocities. In principle, the only limit on the specific impulse achievable with ion thrusters is the operating voltage and the power supply. Hall thrusters use magnetic fields to ionize the propellant gas and to create a net axial-electric field accelerating ions in the thrust direction. MPD thrusters use either steady-state or pulsed-electromagnetic fields to accelerate plasma in the thrust direction. To get a high thrust density, ion thrusters use xenon, while Hall and MPD thrusters can work well with argon or hydrogen. One advantage of electric-propulsion is the higher specific impulses of the electric propulsion. In contrast to a chemical rocket or a nuclear-thermal rocket, they can work continuously for days, weeks, and even months. The main problem is that they have a low power-to-weight ratio and a low thrust density. This makes nuclear electric propulsion infeasible for missions where high accelerations are required.

The drawback of nuclear propulsion is that it is not a fully mature technology. It has a lower technology readiness (TRL) than chemical propulsion. Nuclear thermal propulsion has not such a low TRL. The tests performed in the 1970s on prototype nuclear thermal contributed to the development now of safe, reliable thrusters in reasonable time and with reasonable costs. TRL of nuclear electric propulsion can be split into two parts: the design of the power plant and the design of the thrusters. In the former case, the TRL is advanced. In the latter case, TRL is less advanced. But research aimed at scaling them up and improving their performance is under way. But an important parameter of any nuclear based propulsion system is the specific power that can be achieved. If the nuclear reactor is very heavy, i.e. 100 tons or more, then, even if the mass of propellant is negligible, it may not be competitive with architectures based on chemical propulsion system. The problem is therefore not only to build a nuclear based propulsion system, but to build a light and still very powerful (e.g., 10 MW) one. For that reason, even if the TRL of thermal nuclear based propulsion is not too low, the present TRL of very light nuclear reactors is very low.

Another disadvantage of a nuclear base spacecraft is that it cannot be launched from earth for safety reasons. It has to be sent to orbit before it starts its engine. Atmospheric drag and the risk of re-entry make it difficult to send a nuclear engine at low altitudes. Therefore, for safety reasons, if any nuclear engine had to be used, it should be sent to a higher orbit before its start. The altitude is yet to be defined but would probably be above 1,000 kilometers for a nuclear based propulsion system, while it is around 400 kilometers for a chemical one. The direct consequence is propellant mass penalties for the launches of the nuclear modules and their payloads. These costs have to be added to the development and construction costs of a nuclear based spacecraft. Alternatively, if the new space station foreseen in the previous chapters is in place, the nuclear module and the payload can be sent there to be assembled. The advantage of the assembly in orbit option is the relative independence of the launch rate from the Mars transfer windows[21]. But the costs of lifting the nuclear module and the payload may be quite high, which can be partly offset by strong savings in propellant for the interplanetary travel.

Although the preferred propulsion in our scenario is nuclear, many current studies are considering chemical propulsion. This has implications for insertion in Mars orbit. Higher Δv allowed by nuclear propulsion at injection would shorten the transfer time, but would also result in higher velocities at arrival, which has an impact on the complexity of the Mars orbit injection phase. In this case, at the end of the interplanetary travel, a Mars orbit insertion maneuver is required. There are three options for Mars orbit insertion: (1) all propulsive, a propulsion system with a high specific impulse is used to slow down the vehicle; (2) aerocapture, the vehicle is sent to the atmosphere of Mars for an intense atmospheric braking that requires a heat shield; and (3) aerobraking, same as aerocapture, but there are several passages in the upper levels of the Martian atmosphere to complete the braking phase. In the literature, the third option is generally eliminated because it is time consuming and does not bring important advantages compared to aerocapture. The choice for Mars Orbit insertion is strictly dependent on that on propulsion. If chemical propulsion is chosen, aerocapture must be adopted for all vehicles.

---

[21]To go to another planet using the simple low-energy Hohmann transfer orbit, if eccentricity of orbits is not a factor, launch windows are periodic according to the synodic period. For example, in the case of Mars, the period is 2.135 years or 780 days.

In case of nuclear electric propulsion a propulsive capture maneuver is always assumed. In case of nuclear thermal propulsion, different strategies exist. It is possible to use a mixed strategy: propulsive braking for crew vehicles and aerocapture for cargo vehicles.

Aerocapture is a risky maneuver because the vehicle has to enter a thin corridor. In addition, if the arriving velocity is high, if the vehicle is big or if the shape is complex, it is difficult to protect it and control the attitude during the atmospheric drag. It requires backup strategies if aerocapture fails or if an abort is necessary (for instance if there is a problem with attitude control or thermal protection). The minimum backup strategy could be to be able to abort aerocapture and to proceed to a free return to Earth. An alternative strategy would be that a manned vehicle should have a backup wet propulsion stage to come up with the eventuality of a problem before or during aerocapture.

In summary, nuclear propulsion is more efficient than chemical propulsion. It offers the choice of faster velocities and lower exposure risks to cosmic radiation. But the propellant mass penalties for injection on the interplanetary trajectory are lower for chemical propulsion than for nuclear propulsion. Considerations of risks and propellant penalties of insertion in Mars orbit should thus be given careful considerations in the choice of the propulsion system. Chemical propulsion has lower development costs due to its high TRL, particularly with respect to very light nuclear reactors. But, if many human missions are planned, the nuclear propulsion development costs can be considered as an investment repaid by lower costs of the next missions.

The choice of the propulsion is thus a complex one, involving many parameters. It will depend particularly on the timing and costs of development of the nuclear technology, and the actual decline of costs to escape Earth gravity. It could very well be that the use of nuclear based propulsion systems would come in later on for regular manned Mars missions once these technologies have been developed, while the first human exploratory missions to Mars would likely use chemical propulsion systems[22].

6.2.2 Trajectories

A round-trip mission to Mars is a double rendezvous problem, where trajectory solutions must be found for both the outbound leg from Earth to Mars and the return leg from Mars to Earth. Within a planets' sphere of influence (SOI), the planet acts as the primary attracting body, and outside the planet's SOI,

[22]There exist several mission architectures based on chemical propulsion systems. See Zubrin (1993) and, Salotti (2013).

the Sun acts as the primary attracting body. Trajectories are solved using two-body equations of motion. In the case of a Mars mission, the first segment is the hyperbolic escape from Earth, in which the spacecraft is considered to be in the Earth's SOI. During second segment, the spacecraft is in an elliptical transfer about the Sun. The third segment, the hyperbolic arrival at Mars, is in the Mars SOI. This section focuses on the second segment. Due to the planetary motions of Mars and Earth, the selection of an outbound transfer trajectory to Mars will certainly play a part in the determination of the available inbound trajectories and also dictate the transit times and Mars surface stay.

Two types of ballistic trajectories for human missions to Mars are available: conjunction and opposition class trajectories.

a. Conjunction trajectory concept

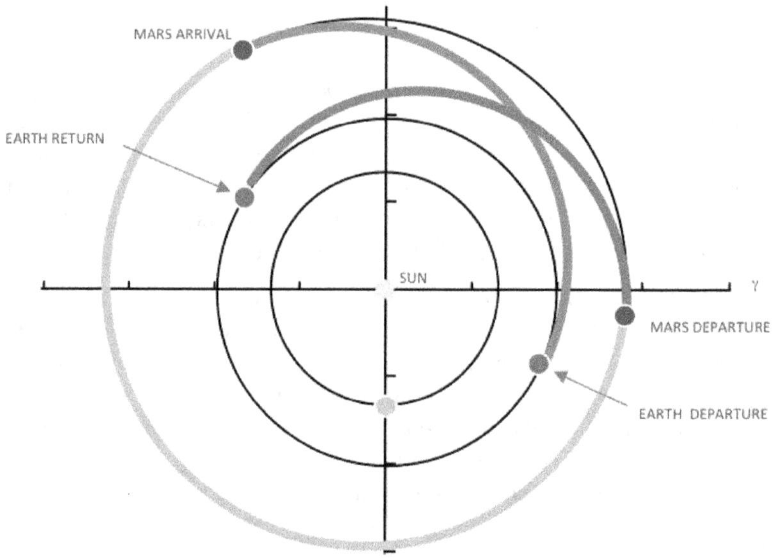

b. Opposition trajectory concept

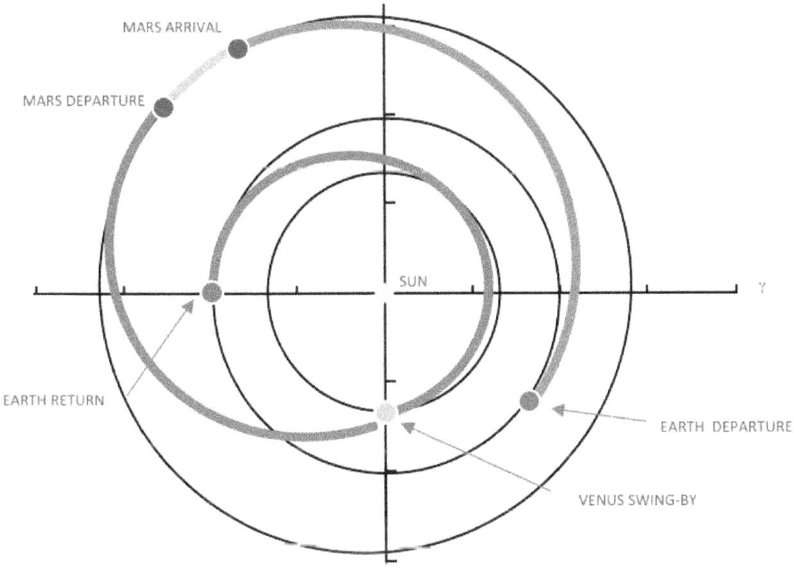

Figure 6.2 – Trajectories for conjunction and opposition Mars missions
Source: NASA

The trade-off between the two options has already been discussed in several papers[23]. A summary of the arguments is presented here. The first option is a minimum energy transfer, which is often called "conjunction class" due to Mars' location behind the Sun as seen from the Earth. They referred to as "long stay" missions, benefit from the planetary alignment of Earth and Mars for both the departure and return legs of the mission, minimizing total Δv requirements. This results in long stay duration on Mars short transit duration, and lower propellant needs. Stay duration on Mars would be on the order of 500 to 600 days. This long stay duration is due to the fact that one has to wait until the respective positions of Mars and Earth are appropriate for a return that is not too much propellant demanding. Total mission duration for conjunction-class missions depends on the propulsion adopted. With chemical propulsion, total mission duration is typically on the order of 900 days

[23]See Zubrin (1990), George et al. (1998), Landau et al. (2006), NASA (1997), and DRA5 (2009).

(400 days total transit time), while with nuclear propulsion the total mission duration is reduced to about 840 days (300 days total transit time).

The second is a high energy transfer called "opposition class" due to Mars' location on the opposite side of the Earth from the Sun. It is one in which the stay on Mars is short and the interplanetary travel is longer. In the reviewed literature, the stay on Mars is typically about thirty to forty days. With chemical propulsion, the total interplanetary transit time is about 500 days (395 days in the conjunction trajectory). The major difference is due to a longer inbound transit time. With nuclear propulsion, the total travel time could be reduced, but at the cost of a shorter stay on Mars. The short stay on Mars is due to the fact that departure from Earth should generally be performed earlier than in conjunction trajectories in order to rendezvous with Mars a few weeks or months before the conjunction of the two planets and to be able to return to Earth.

Chart 6.1 - Comparison of three mission profile (days)

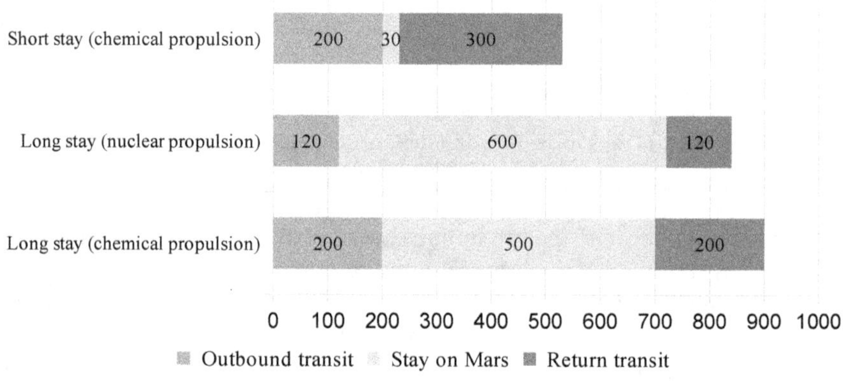

Credit: author

Evaluation of the propulsive requirements for both trajectory concepts yields substantially larger total Δv requirements for opposition trajectories. These impacts of Δv on propellant requirements can be demonstrated by the ideal rocket equation (1), which yields exponential growth in propellant as Δv increases. Equation (1) can be rewritten as:

$$m_{pr} = m_1 * e^{(\Delta v / ve)} - m_1 \quad (2)$$

where m_{pr} is the required propellant mass.

This equation indicates that, given a fixed propelled mass and exhaust velocity, a 100% increase in Δv will result in an almost 400% increase in propellant mass. Total Δv for a conjugation trajectory is typically estimated at 2.81 kilometers/second, while for an opposition trajectory is 5.69 kilometers/second, implying an increase in Δv of 103%. A Venus swing-by may eventually help to reduce the required amount of propellant, but it would still be much more demanding than for the conjunction trajectory.

Important assessment criteria in choosing the interplanetary trajectory are: efficiency in terms of propellant's consumption; the transit time; astronauts' safety; and the stay time on Mars. Efficiency in terms of propellant's consumption favor the adoption of a conjunction trajectory. As the dangers to the crew, apart from the launch, comes mostly from the time spent in space, owing to their exposure to radiation and to the long periods of weightlessness, transit time and astronauts' safety considerations favor conjunction trajectories. Moreover, dangers due to radiation are further increased in the inner solar system approaching the orbit of Venus for the flyby. Time spent on Mars is crucial to the mission's goals. The adoption of an opposition trajectory would limit the time spent on Mars. This limits surface productivity by denying the astronauts of adequate time to: develop significant Mars infrastructure, demonstrate the feasibility of sustained in-situ resource utilization, thoroughly investigate the terrain around the landing site, and assess the effects of low gravity on the human body in long duration stay.

The conjunction trajectory has the advantage of low energy requirements. This trajectory provides an opportunity to send a more massive spacecraft (i.e. more cargo) at the same cost of a smaller spacecraft following an opposition trajectory. By maximizing the payload of each launch vehicle, the number of launches necessary to transport the required surface equipment would be minimized. The disadvantage of this trajectory is the longer exposition of astronauts to radiations on Mars. Risks to the astronauts to in space radiations are lower than the ones following an opposition trajectory, but still high, although not as much shielding would be required since the spacecraft would never be inside of Earth's orbit. In summary, the main mission goals (efficiency, and astronauts' safety) rule in favor of conjunction-class trajectories.

6.2.3 Shielding of cosmic radiation

There are two types of radiation that need to be addressed for long duration human space flight. The first originate from solar flares, which flare eject into space clouds of electrons, ions, and atoms through the Sun's corona. They produce streams of highly energetic particles in the solar wind, known as a solar proton event. The frequency of occurrence of solar flares varies. It ranges from several per day when the Sun is "active" to less than one every week when the Sun is "quiet". The second source is galactic cosmic rays, which, although less lethal than solar flares, constitute continuous radiation to which the crew is exposed. Exact measurements of the cosmic-ray environment are yet needed to plan proper countermeasures[24].

Several strategies are being studied to protect astronauts the radiation hazard. One strategy (passive absorbing shielding) is to build the spaceship with hydrogen-rich plastics such as polyethylene-based material. Hydrogen is good for shielding against cosmic rays. The plastic materials' advantage is that they produce far less secondary radiation[25] than heavier materials such as aluminum or lead. Studies are progressing at NASA to develop polyethylene-based materials that combine superior structural properties with superior shielding properties. But shielding with hydrogen-rich plastic has a limit. If too much shielding material is used, this affects the efficiency of the mission. But a way out exists: limiting the mass of shielding material can be achieved through wise scheduling of the crew's activities, since significant amounts of crew time could be spent in an on-board sheltered area during the cruise phase. This limits the shielding-material mass while enhancing the safety of the crew. A second strategy is to surround the spaceship with a magnetic field. This strategy has already been discussed in Chapter 5 with respect to shield astronauts in the new international space station. As the technology should have been already developed at the time of the Mars mission, this second option should be the preferred one.

A third strategy centers on increasing astronauts' resistance to the deleterious consequences of radiation exposure. A study commissioned by ESA[26] is investigating: the potential use by crew members of prophylactic radio-protective drugs; and possibilities to control phytochemical antioxidants in the body through dietary choices. Scientists are looking for ways to help the body after the damage is done. One researcher is exploring the cell cycle. As a cell divides, it pauses occasionally, to check its genes for any damage and to repair errors. With pharmaceuticals that lengthen this part of the cycle, researchers believe they can give the cell more of a chance to fix its own problems. Investigations of the biological response of the body to energetic particle radiation under conditions of micro-gravity are underway through dedicated animal space flight experiments.

---

[24]A physics experiment on the International Space Station will help in these measurements. The Alpha Magnetic Spectrometer (AMS) will be attached outside the station. It will search for different cosmic rays, measuring their long-term variation over a wide energy range, and for nuclei from protons to iron.

[25]Secondary radiation comes from the shielding material itself. When particles of space radiation smash into atoms within the shield, they trigger tiny nuclear reactions. Those reactions produce a shower of nuclear by-products - neutrons and other particles - that then enter the spaceship.

[26]See Horneck, G. et al. (2003).

6.2.4 *Descent/ascent vehicles*

Two issues are analysed here: the descent vehicle for cargo and astronauts, and the ascent vehicle for astronauts to leave Mars surface. The descent to the surface is one of the most complex problems of the mission. Several parameters have to be considered: the size and mass of the payloads that have to be landed on Mars; the ballistic coefficient of candidate landing vehicles; the initial parking orbit; and the altitude of the landing zone. According to recent studies[27], rigid deploy-able heat shields or hyper-sonic inflatable atmospheric decelerators (HIAD) are promising systems.

Rigid deploy-able heat shields have already been tested on Mars. The Mars Science Laboratory (MSL) spacecraft that landed the Curiosity rover on Mars endured the hottest, most turbulent atmospheric entry ever attempted in a mission to the Red Planet. The saucer-shaped MSL was protected by a thin, lightweight carbon fiber-based heat-shield material (PICA) that was a bit denser than balsa wood. While traditional heat shields form a rigid structure, NASA Ames Research Center, which developed the PICA material, is now developing a new family of flexible heat-shield systems that use a woven carbon-fiber substrate, or base material. This material's heat-resistance and structural properties can be fine-tuned by adjusting the weaving techniques. In addition, the flexible nature of woven materials can accommodate the design of large-profile spacecraft capable of landing heavy payloads (less than forty tons) required for a human mission to Mars.

The Hyper-sonic Inflatable Aerodynamic Decelerator (HIAD) design consists of an inflatable structure that maintains the aeroshell shape against the drag forces, and a protective flexible thermal protection system that withstands the thermal loading. For destinations with a sensible atmosphere, aerodynamics (specifically atmospheric drag) provides the most mass effective way to decelerate a payload to a soft landing. Hyper-sonic vehicles entering the atmospheres of planets are traveling so fast that they create a high energy pressure wave. This pressure wave entraps and rapidly compresses atmospheric gases, resulting in drag forces that decelerate the vehicle and intense thermal loads that heat the vehicle. The HIAD overcomes size and weight limitations of current rigid systems by utilizing inflatable soft goods materials that can be packed into a small volume and deployed to form a large heat shield before atmospheric entry. Since the beginning of its development

[27]See Salotti (2015).

in 2003, several ground based advancements and two successful flight tests[28] have provided a team of engineers at NASA Langley Research Center valuable data used to refine the HIAD technology. HIAD project is developing and demonstrating an enabling technology that will accommodate the atmospheric entry of heavy payloads (over forty tons) to planetary bodies such as Mars. However, the large HIAD vehicle may have steering issues during the guided phase of flight.

Different configurations of ascent vehicle are analysed in human Mars mission proposals. They fall into one of two broad categories: a relatively large habitable variant that serves as both ascent vehicle and habitat, either on the surface or in orbit; or a much smaller taxi used only for a few hours during the actual ascent. Ascent vehicles are generally regarded as the largest "gear ratio" item in a given architecture; in other words, every kilogram of an ascent vehicle needs more Earth-launched mass to do its job than most other architecture elements need. Even if ascent propellant can be manufactured on Mars, current technologies still require the ascent vehicle and the propellant manufacturing plant to be launched from Earth, transported to and then descended onto Mars, all of which entails a considerable amount of transit and descent propulsion mass. One promising solution involves a smaller taxi variant with the integration of the descent vehicle for astronauts. This solution allows the same propulsion system to be used for descent to, and ascent from, the surface of Mars.

6.2.5 Landing site

There are four criteria for choosing a landing site: the presence of water; the availability of building materials; safety issues; and, most importantly, the mission's goals. Preliminary scouting for potential sites of a permanent human base weighs heavily in the landing-site choice. Five potential sites seem promising. *Tharsis* is a vast volcanic plateau centered near the equator in Mars' western hemisphere. Tharsis is home to the largest volcanoes in the solar system. They include three enormous shield volcanoes: Arsia Mons, Pavonis Mons, and Ascraeus Mons. Pavonis Mons hosts several lava caves. Images from orbiting vehicles have identified skylights or breaks in the top of buried lava tubes. These can offer shelter from radiation and a constant ambient temperature. This site presents few drawbacks, but access to the lava

---

[28]The Inflatable Re-entry Vehicle Experiment-3, or IRVE-3, suborbital launch of a 300 kilograms re-entry vehicle to a height of 460 kilometers on July 23, 2012, was the most recent HIAD test flight. The IRVE-3 flight test successfully demonstrated the ability of a 3 meters diameter HIAD to decelerate the vehicle from an entry speed of more than 2700 meters/second with drag forces as high as 20 g.

caves is challenging. They are at high elevation on the volcanoes and at discreet locales. Moreover, unlike the northern plains, Mars' lava tubes might not have water.

Arabia Terra is a northern highland with dense impact craters dating back billions of years. Alongside many craters, canyons stretch through Arabia Terra. Many terminate at the northern lowlands bordering Arabia Terra to the north. Craters offer protection from winds, which can be very strong and full of dust. Water, most of it bound within minerals, is present in Arabia Terra. In terms of energy and technology, water will be complicated and costly to extract. But water may also be present inside a crater created by a comet's impact with Martian soil[29]. Images show an iced lake a. few kilometers' in size and with an estimated depth of 100 meters. The major drawback of this landing site is the lack of protection against radiation.

Hellas basin is located in the southern hemisphere. Hellas basin has many glaciers. They are fossil remnants with iced water at shallow depths. But glaciers are fractured and with many crevasses. It would likely be necessary to land off the glacier and traverse over to it to bring the ice and water back to the astronauts' habitation module. Just as on Earth, walking on Martian glaciers would also be dangerous. As in Arabia Terra, no natural shielding exists against radiation. A *fourth site* is composed of the canyons of Valles Marineris, where spring water deposits may exist as it periodically breaks through ice plugs. Daytime temperatures can be around zero degrees Celsius. While surface water may exist only a couple of times a year, the source liquid aquifers may be accessible. But accessing aquifers up slopes hundreds meters above the landing site will not be an easy job.

The last potential site is in the *outer hedge of the North Polar Cap.* Water is close, while this flat region is safe for landing. Mars' North Pole is six kilometers lower in elevation than the South Pole which offers more time to slow landing craft and correct their position. Since Mars' solar orbit is greater than Earth's, its northern summer has nearly six months of continuous daylight. This allows direct-to-Earth communications most of the time, in contrast to non-polar sites, which have to rely on links through orbiters.

6.2.6 The Mars outpost base camp and logistics

The base camp consists of a central habitat, a life science module, and the power system. Several proposals for the habitat's architecture exist. These proposals are categorized into two classifications. The first ones are prefabricated structures deployed on Mars surface: the Transit Habitat or Tran-

---

[29]ESA Mars Express mission identified this crater.

sHub. TransHab is a unique hybrid infrastructure that combines a hard central core with an inflatable exterior shell. The central core is the hard part comprised of longerons, re-positionable launch shelves, bulkheads, radiation shield water tanks, utility chases, and integrated duct work. The inflatable shell is composed of four layers: the internal scuff barrier and pressure bladder, the structural restraint layer, the micrometeorites shield, and the external thermal-protection blanket. An integrated pressurized tunnel is located at one end to enable access to the Mars surface. A tunnel, not pressurized, is located at the opposite end. It houses the TransHab inflation system. Because TransHab is prefabricated, packaged and deployed, it requires to be transported from Earth and the crew to set it up.

The second ones foresee structures manufactured using local resources and have already been explored in section 5.2.2. Its construction requires on the surface a regolith drill/pump, a brick baker, plastic super-adobe bags to hold soil, an inflatable half-cylinder membrane, and a plexiglass sheet supply. The brick baker is a simple device. It pours regolith into mild compression molds. A small furnace bakes the regolith to produce bricks in the mold's shape. The habitat wall is built using coil and bags filled with regolith. Each layer is placed on top of lower layers and hardened by compression of the layers above Specific spaces are built into the wall for power and gas lines, plexiglass windows, and airlock junctions. The roof is built using the "leaning arch" technique. After the exterior is built, the crew will lay the base floor matting. They will bring in the inflatable to match up its airlocks with the wall openings. The membrane is inflated with a mixed oxygen gas[30]. Buffer gases are regulated at nine psi supply pressure. Internal supplies, furniture, and equipment are brought in after inflation. Two inflatable tunnel airlocks are attached to the habitat, one joining the lander and the other joining the life-science module.

Both habitats offer adequate solutions to protect astronauts against wind, radiation, and low temperatures. The Abode infrastructure serves as a shell to house the pressurized membrane. In contrast, TransHub needs hard structural support to mold and contain the nine psi environment. Adobe infrastruc-

---

[30]While is possible for humans to breathe pure oxygen, a pure-oxygen atmosphere is dangerous as demonstrated by fire in the Apollo 1 capsule. So the Mars habitat may need additional gases. One possibility is to take nitrogen and argon from Mars atmosphere. Separating nitrogen from argon is difficult. It is suggested that the Mars habitat uses 40% argon, forty 40% nitrogen, and 20% oxygen. Thanks to modern technology, Mars atmosphere can be compressed and adjusted to form a breathable mixture for humans. But the effects on humans of breathing gaseous argon are unknown. Ongoing experiments are being conducted to monitor the health and activities of different animals while breathing a gas mixture that includes argon.

ture reduces the overall mass transported to Mars and the mission's costs and offers useful information and experience for the permanent human base on Mars. Attaining cost efficiency and long-term goals rule in favor of the Adobe infrastructure.

A study prepared for the Second International Energy Conversion Engineering Conference sponsored by the American Institute of Aeronautics and Astronautics (2004) indicates that an initial temporary Mars settlement, including life support consumables, production of propellants for surface mobility and the ascent vehicle, and settlement construction materials, would require energy between 100 kW and 1000 kW, depending on the size of the base. Several options are being investigated by space agencies and research laboratories, including solar and nuclear power. The preferred option includes a nuclear reactor coupled with a Stirling engine to generate electricity. A liquid metal[31] transfers the heat from the reactor to the Stirling engine, which uses gas pressure to convert heat into electricity. The nuclear reactor should be small, compact, and safe. One candidate is an SP-100 reactor[32]. The reactor has integral shielding and is self-contained. It includes its own auxiliary power for start up. Components including the reactor, primary coolant, and power conversion components are in a vacuum vessel for protection from the CO_2 environment. The vacuum is maintained by an active-ion pumping system. These active-ion vacuum systems have no moving parts and have a long record of reliable operation. Assuming a specific mass value of 150 kilograms per kilowatt hour, the reactor mass is thirty tonnes.

The rover-enabled exploration focuses on expanding our knowledge of the Martian environment while searching for a potential location for the permanent base camp. Exploration is performed using a long-range traverse vehicle. This vehicle will be capable of traversing the Martian landscape while providing a comfortable living and working environment. The vehicle is supplied by cargo lander that will place supply caches at several locations along the traverse route. Each cache provides human necessities along with scientific and monitoring equipment. A rover vehicle with thirty kW user net power is being considered. The operations envisaged as enabled by this power include: drilling, sampling and sample analysis, onboard computer and computer instrumentation, vehicle thermal management, and astronaut life support. Besides

---

[31]The liquid metal is a sodium potassium mixture. It is used to transfer heat from the reactor to a generator.

[32]SP-100 (Space Reactor Prototype) is a U.S. research program for nuclear-fission reactors in space. A significant part of the SP-100 Project is aimed at identifying plausible missions and determining their requirements. Preliminary results indicate significant benefits of nuclear-reactor power systems.

the thirty kW net user power, electric power is needed to propel the rover across the Martian terrain.

The only system that can offer this level of power at reasonable mass penalty is a nuclear-reactor heat source with static or dynamic conversion. Current research efforts are focused on the NASA SP-100 reactor. To integrate a nuclear reactor into the long-range traverse vehicle, one needs a detailed description of the rover's requirements. A pressurized rover may have the size of a city bus with a mass of around eight tonnes. The vehicle performance requirements are: a velocity of ten km/hr; a range of hundred kilometers; a slope climbing ability of $^{o}30$ uphill for fifty kilometers; a single mission duration of five days; and a crew of two. A variety of technical issues need to be resolved. These issues include: minimization of the shield mass for protection of the crew from nuclear radiation; optimization of shield configuration; and integration of the heat emitting radiator with the rovers design. The power system's specific mass of 100 kilograms per kWh yields a total vehicular mass of ten tons, including three tonnes for the reactor based on SP-100 technology. The remaining seven tonnes of mass is devoted to the radiation shield. The vehicle will have an auxiliary power system (a radioisotope power system[33]) as a back up in case of failure of the primary one.

One further power source, although relatively minor, is needed. Astronauts need spacesuits to work on Mars; the ideal spacesuit is an exoskeleton power-assisted, pressurized one. It is well-insulated but light-weight and flexible to accommodate the astronauts' movements. It is different from the bulky, unwieldy ones used by the Apollo astronauts. Space agencies are continuing to study this new spacesuit. An important problem waiting solution is the exoskeleton's energy source. The most promising power solution centers on new electricity accumulators based on nano-capacitors[34]. Nano-capacitors are formed by clusters of platinum atoms surrounded by clouds of carbon atoms. Among the current studies is an interesting experiment on electric-charge accumulation being conducted in Italy at the University of Milan.

---

[33]Radioisotope power systems provide electricity and heat enabling spaceships to undertake missions to environments beyond the capabilities of solar power, chemical batteries and fuel cells. These technologies are capable of producing electricity and heat for decades under the harsh conditions of deep space without refueling.

[34]Nano-capacitators store electrical charge on two metal-electrode surfaces separated by insulating material. Their storage capacity is directly proportional to the electrodes' surface area. Electrodes work in the same way as in conventional capacitors, but, instead of being flat, they are tubular and deep inside the nanopores.

6.3 Conclusions

Preliminary human exploration missions are necessary before humanity embarks on building a permanent human base on Mars. Current proposals to send humans to Mars in a one way mission, without necessary preliminary exploration and verification in situ of the conditions for living on Mars, are too risky and may result in a set back of the development of the space economy beyond the Moon. These preliminary missions can be carried out only by space agencies. Sending humans to explore Mars, giving them all that is needed to perform their tasks on the planet, and bringing them safely home is a formidable task, which requires large financial resources and a long term engagement.

Estimated total costs of a manned exploratory mission to Mars range from hundred billion dollars to one trillion dollars. The figure of one trillion dollars was suggested by the media when President Bush unveiled a new space policy calling for a human exploration mission to Mars. Regardless that this number reverberated within the modern media echo chamber, no detailed study supported that number. Hence the hundred billion dollars is taken as a reference cost. The most advanced studies are those connected with NASA's Design Reference Mission (DRM). NASA issued a series of DRM versions between 1993 and 2009. These missions are characterized by a small crew and several small interplanetary vehicles launched in one piece and powered by chemical propulsion. The surface explorations, solely striving to achieve scientific goals, are limited.

The mission outlined in paragraphs 6.2 differs in several respects from NASA's Design Reference Mission. The manned spaceship is powered by nuclear propulsion while the cargo spaceships are powered by either chemical or nuclear propulsion. The scope of Mars surface exploration is wider, and it involves a heavier and more sophisticated long-range traverse vehicle. The habitat structures are manufactured using local resources. In contrast, the habitat in NASA's Design Reference Missions (DRM) comprises pre-fabricated structures deployed on Mars surface. The base camp's power supply relies on a nuclear reactor, considered safer for long-term stay missions. Hence the costs of this mission would be larger than the ones estimated in the various DRM missions proposals. As no single nation can afford these costs, International collaboration would be required. Although international collaboration would reduce each space agency funding of the project, it is likely that participating space agencies' budgets should be increased to finance these missions. A more stable world resulting from space mining could generate these financial resources.

Planning and executing an exploratory mission to Mars would require long term commitments by space agencies. These commitments should not be left to the ebb and flow of national policies, but should be enshrined in an international agreement among space agencies with the eventual collaboration of international space institutions. A proposal would be that a space-agencies' consortium under United Nations aegis will carry out Mars exploration. This consortium will be open to every space faring nation. Investment quotas could be calculated according to the ratio of each space agency's budget in the total budget of participating space agencies. The consortium would be responsible for: designing the mission and coordinating space agencies and industries activities. This would allow to program in advance inter-operability through the strategic use of redundancy. In case of default of any one of the main partners, the maintenance of sufficient production capacity by other main partners would compensate for any lost skills relatively easily. Partners without critical technologies would reach agreements with main partners should they not have adequate production capacity or when non-critical partners can supply technologies at lower costs.

If Mars exploration is accomplished during the 2030s, a permanent base on Mars could be built at the end of the 2030s or the beginning of 2040s. This would give enough time to private sector proposals, such as Elon Musk's, to refine their concepts and, more importantly, to reduce costs of traveling of the settlers to Mars. The success of Mars colonization depends heavily on the political will of governments to pursue rapidly the construction of a cislunar space station and to a long term commitment for the human exploratory mission to Mars based on international collaboration. Obstacles to these endeavors are thus not technical but political.

Bibliography

Billington R. A.: *"America's frontier heritage"*, ed. Holt, Rinehart and Winston, New York, 1966.

Borowski S.: *"Human Exploration of Mars. Design Reference Architecture 5"*, NASA-SP-2009-566, 2009.

Dyson F. J.: *"Disturbing the Universe"*, Basic Books, N. Y., 1979.

ESA: *"Aurora's road map to Mars / Exploration / Human Spaceflight / Our Activities / ESA"*, European Space Agency, 2003-12-19.

George L. E. et al.: *"Interplanetary mission design handbook: Earth to Mars mission opportunities and Mars to Earth return opportunities 2009-2024"*, NASA-TM-1998-208533, 1998.

Landau D. F. et al.: *"Trajectories for Human Missions to Mars, Part 1: Impulsive Transfers"*, Journal of Spacecraft and Rockets, Vol. 43, No. 5, September–October 2006.

Mars Architecture Steering Group: *"Human Exploration of Mars Design Reference Architecture 5.0"*, NASA/SP–2009–566, 2009.

Mars Architecture Steering Group: *"Reference Mission Version 5.0 Addendum"*, NASA/SP–2009–566-ADD, NASA, 2014.

Musk E.: *"Making Humans a Multi-Planetary Species"*, New Space, 5(2): 46-61, June 2017.

NASA: *"Report of the 90-Day Study on Human Exploration of the Moon Mars"*, NASA-TM-IOZ_12999, 1989.

NASA: Scholastic, Teacher Activity Guide, *Buzz Aldrin Interview*, November 17, 1998, http://teacher.scholastic.com/space/apollo11/interview.htm

NASA: *"Vision for Space Exploration"*, 2004.

Ordway III F. I. et al.: *"EMPIRE: Early Manned Planetary-Interplanetary Round trip Expeditions Part I: Aeronutronic and General Dynamics Studies"*, Journal of the British Interplanetary Society, May 1993, pp. 179-190.

Salotti J. M.: *"Revised Scenario for Human Missions to Mars"*, Acta Astronautica, vol. 81, p. 273-287, 2012.

Salotti J. M.: *"Revisiting Mars Semi-direct"*, paper presented at the 9th IAA Symposium on the future of space exploration: towards new global programs, Turin, July 2015.

von Braun Wernher W.: *"Project Mars: A Technical Tale"*, Apogee Books, 2009.

Weistein L. M.: *"Space Colonization using Space Elevators from Phobos"*, paper developed at the Advanced Measurement and Diagnostics Branch, NASA Langley Research Center, 2003.

Zubrin R. et al.: *"Mars Direct: Humans to the Red Planet by 1999"*, Proceedings of the 41st Congress of the International Astronautical Federation, 1990.

Zubrin R. et al.: *"Practical Methods for Near-Term Piloted Mars Missions"*, AIAA 93-2089 AIAA/SAE, 29th Joint Propulsion Conference, Monterrey CA, 1993.

Chapter 7

The ultimate challenge: the exploration and colonization of extrasolar planets.

"I have learned to use the word impossible with the greatest caution."

Wernher von Braun

The previous chapters have dealt with the development of the space economy in the short- medium-long term. In those cases, most technologies to achieve missions' goals are mature and available. Major show-stoppers are institutional, political, and economical. This chapter deals with the longer term development of the space economy, i.e. interstellar travel and colonization. Given the enormous distances among the stars[1], some consider interstellar travel practically impossible. Impossible is, however, a relative term. It depends on the knowledge or lack of it of basic laws of physics. This chapter makes the case that, if an extrasolar planet capable of sustaining life is discovered within fifteen light years from Earth, the knowledge of the basic law of physics exists, allowing such travels. But they require the development of technologies that are not yet mature. To go further than fifteen light years from Earth is, at the moment, impossible since we do not have the scientific knowledge or the technologies to accomplish these missions.

7.1 How to go there: Einstein and the special relativity theory

Interstellar distances are large. This restricts the mission's targets within ten/fifteen light years from Earth. To reach these stars in reasonable time, it is necessary to achieve velocities equal to significant fractions of the speed of light. To attain such speed and shed it in reaching a destination will pose formidable challenges to spaceship designers. When approaching the speed of light, the spaceship must use more energy to achieve the same acceleration as at lower speeds. In theory, nuclear-fusion propulsion can run with continuous acceleration and achieve significant fractions of the light's speed. But

---

[1]The two Voyagers interstellar spaceship - the fastest ever launched from Earth - are traveling at one-thousand[th] of the speed of light. They would need about 70,000 years to go to the nearest star, making human interstellar travel looks unrealistic.

the mass of fuel required is large. Not to bury the reader with complicated calculations, the mass of fuel should be a multiple of the spaceship's mass. The conclusion is that interstellar travel with nuclear-fusion propulsion, although possible, is not practical. Someone (i.e., Bussard[2]) suggested a way out, namely, to collect more fuel during the trip. Hydrogen is the most abundant material in the universe, and the average density of hydrogen in our galaxy is one hydrogen atom per cubic centimeter. To gather one gram of hydrogen per second, the spaceship should use collection panels more than forty kilometers wide, another impractical solution.

To solve the fuel mass problem, the recipe is more efficient fuel with higher energy content per unit mass. Annihilation of particles with their antimatter counterparts has the highest ratio known in physics between energy produced and mass used. When matter and antimatter meet, both "disappear", resulting in energy according to Einstein's equation $E = mc^2$. Multiplication by such a large number as the square of the light's speed, shows that a small mass may give rise to a great deal of energy. Articles in the technical literature[3] have analysed antimatter as an energy source for interstellar vehicles. The only engine that can reach 50-60 per cent of the speed of light is one that uses products of the of matter-antimatter annihilation directly to generate thrust. This system is called the pion engine. Pions are produced by proton-antiproton annihilation. The resulting charged pions will have a velocity of 94 per cent of the speed of light. A Lorentz factor of 2.93 extends pion's lifespan enough to travel 2.6 meters through the nozzle before decaying into muons. Sixty percent of pions will have either a negative or a positive electric charge. Forty percent of pions will be neutral. Neutral pions decay into gamma rays, useless for generating thrust. In realistic matter/antimatter reactions, collimation may not be perfect, and few pions are deflected backwards by the nozzle. Thus, effective exhaust velocity for the entire reaction drops to between 50-60 per cent of the speed of light.

The antimatter mass needed to power engines for interstellar travel is large. It ranges from few kilograms to a few tonnes, depending on spaceship mass and cargo. Two problems exist. Antimatter does not exist in the observable

---

[2]See Bussard R.W. (1960).

[3]See Lewis, R.A., et al. (1997); and Frisbee R. (2003).

universe[4]. We have to produce it. So one main problem is how to produce antimatter and at what cost. At CERN in Geneva, particle accelerators produce 1015 antiprotons per year. If this seems like a large number, to make one gram of antimatter takes 1026 antiprotons! At this rate, production of one gram will take centuries, not to mention what would be required to produce kilograms or tonnes. Moreover, to produce antiprotons needs energy. Unfortunately, the energy needed is greater than the one obtained by converting mass into energy. This raises the cost of producing one gram of antimatter to astronomical dimensions levels. G. R. Schmidt[5] has proposed a way to calculate the cost of antimatter production. Antimatter' production is energy-intensive. Converting input energy (E_{in}) into the energy (E_{out}) of the collected antiprotons' mass can be expressed in terms of efficiency:

$$\eta = E_{out} / E_{in} \qquad (1)$$

Because of the equivalence between E_{out} and the antiproton mass M_a ($E_{out} = M_a c^2$), the energy required to create a unit mass of antiproton is given by:

$$E_{in} / M_a = c^2 / \eta \qquad (2)$$

The energy needed to produce antiprotons is inversely proportional to the conversion efficiency factor (η) as the speed of light is constant. The total cost of energy (C) to produce antiprotons is then obtained by multiplying the equation (2) by grid energy cost (C_{grid}) and by M_a:

$$C = C_{grid} M_a c^2 / \eta \qquad (3)$$

In present particle accelerators, η is around 10^{-08}. Assuming C_{grid} around 0.1 dollar per kilowatt hour, this translates into an energy cost of around 63 trillion dollars per gram of antiprotons. So we are still far from producing a cost-efficient fuel to power a matter-antimatter engine. If the cost of antimatter is brought down to ten million dollars per milligram, matter-antimatter propulsion will be less expensive. It will be competitive with nuclear-fusion propulsion, depending on the efficiency of the design of the various engines.

In the half century since the antiproton discovery, the intensity of antiprotons beams has increased by at least a factor of 10^4. The increase was not linear but exponential. This is well documented. Who knows what accelerator physicists will be able to do to increase efficiency of antiprotons' production

[4]Signs of antimatter's presence in the universe comes through annihilation of matter and antimatter. If these two particles coexist, we can expect to see characteristic radiations emitted in annihilation. No sign of these radiations is detected. Radiation detected is generated by collision of high-energy cosmic rays with each other or with Earth's atmosphere. Small antimatter amounts are produced in these collisions. These results show that antimatter exists in secondary form and in small amounts in the universe.

[5]See Schmidt G. R. and al. (1998).

in the next half century. Current particle accelerators produce antiprotons only for scientific experiments, not the most efficient way. Few visionaries are imagining the construction of accelerators dedicated to the production of antiprotons, which would decrease power losses during production and increase efficiency. Robert Forward is one such visionary. Forward[6] analysed what happens when a particle accelerator is designed for no other purpose than to create antiprotons. Results show energy-efficiency increases from one part in 60 million to one part in 10,000. This important improvement approaches the "break-even" point compared to nuclear-fusion engines.

Once efficient ways to produce antimatter are found, the remaining problem is to conserve antiprotons. At Fermilab and CERN, they do it with special magnetic bottles known as Penning traps[7]. Results have been modest. Experiments show that one billion antiprotons can be stored for one year. This is far from the antimatter mass needed by interstellar spaceships but is a start. A century ago few visionaries[8] imagined using liquid propellant for rockets. Even though it was a big step forward from solid propellant, scientists and many commentators mocked them. They claimed that producing liquid hydrogen and using it in an engine was too costly and dangerous. Half a century later, current use of liquid hydrogen had become a reality and its costs are reduced. More than half a century after the first microdroplets of liquid hydrogen were produced in a Dutch laboratory, liquid hydrogen sent men to the Moon.

The methods described above make it possible to produce small quantities of antimatter at a reasonable cost. Spaceships powered by matter-antimatter propulsion can then be built. The engines will use energy from the annihila-

---

[6]See Forward R. (1985).

[7]In vacuum, the particles of antimatter can be trapped and cooled within a magnetic trap using a highly concentrated laser beam. Another method, more hypothetical, is the storage of antiprotons inside fullerenes. In theory, the negatively charged antiprotons can repel the electron cloud around the nucleus of carbon. Antiprotons will not get close enough to the protons of the nuclei and avoid annihilation.

[8]The idea of a liquid rocket as understood in the modern context first appeared in the book "*The Exploration of Cosmic Space by Means of Reaction Devices*" by K. Tsiolkovsky. During the 19th century, the only known developer of liquid-propellant rocket engine experiments was Peruvian scientist Pedro Paulet.

tion to heat a propellant for thrust[9]. The main differences between these engines are in their performance characteristics: exhaust velocity, energy efficiency, and mass ratios. For example, using ACMF propulsion, the exhaust velocity can reach 100 kilometers per second. Using AIM propulsion, the exhaust velocity can reach 1000 kilometers per second. Antimatter needed for a one-year round trip to Jupiter by a spaceship powered by ACMF engine is estimated at around ten micro-grams (μg). Antimatter needed by a spaceship powered by AIM engine for a 50-year round trip to the Oort cloud is around 100 μg.

Antimatter needed for interstellar travel will be instead in the order of kilograms or tonnes. To produce these amounts requires technologies that go beyond those already described. Again, Forward in his book *Indistinguishable from Magic*[10] anticipated a possible solution. He proposes to move antimatter production from Earth to space. Here is what he has to say:

> "Where can we get the energy needed for the large-scale production of antimatter? Some of the prototypes will be built on Earth, but for the large scale production we do not want to feed these machines with the combustion of fossil fuels on Earth. There is plenty of energy in space. At the distance of the Earth from the Sun, the Sun provides more than a kilowatt of energy per square meter of collector or a gigawatt per square kilometer. A collector field of 100 square km would provide a power of ten terawatt, which is sufficient to produce one gram of antimatter per day."

Remember that in half a century asteroid mining will be in full swing. Raw materials from space will be available to manufacture space infrastructure at reduced costs. It would be plausible to produce antimatter in space at reasonable costs, opening the way to interstellar traveling.

---

[9]A proposed engine is called Antimatter-Catalyzed Micro-Fission/Fusion (ACMF). In this engine, a pellet of deuterium-tritium and uranium 238 is compressed by particle beams and irradiated with a low-intensity beam of antiprotons. The antiprotons are absorbed by the uranium 238. This starts fission that heats and ignites the deuterium/tritium core, which expands and produces the pulsed thrust. A second proposed engine is called Antimatter-Initiated Microfusion (AIM). In this engine, antiproton plasma is compressed by electric and magnetic fields, while droplets of deuterium/tritium are injected into the plasma. The antiprotons annihilate on contact with the deuterium/tritium droplets, which heat the plasma to ignition conditions. The products of this ignition are directed to a magnetic nozzle to produce thrust.

[10]See Forward R. (1995).

7.2 The search for extrasolar planets capable of sustaining life

A first gauge of a planet's habitability is its orbit around its sun, the so called habitable zone. One may define this orbit as one where a planet receives enough solar irradiation, indicating the possible presence of liquid water on the planet's surface. A second gauge is the existence of a planet's atmosphere and its composition. Atmospheric composition is a significant indicator of a planet's habitability. Oxygen and nitrogen are important chemical elements for life. Oxygen is an unstable chemical element. Its presence in the planet's atmosphere is a good indicator of the presence of external agents keeping the atmosphere in disequilibrium. We know that the process of photosynthesis, performed by plants or cynobacteria, creates the conditions for disequilibrium in Earth's atmosphere. But we cannot rule out other unknown agents in extra-solar planets.

A third gauge is the presence of a magnetic field around the planet. The absence of a magnetic field would leave the planet at the mercy of fast moving electrically charged particles that are blown towards the planet by its star. This would gradually blow away any atmosphere around the planet, leaving in the end about nothing, as happened to Mars. Although the planet may be habitable in the short term, in the longer term it would become difficult to inhabit by human beings. Another important gauge is the presence of a moon. On Earth, the presence of the Moon has a powerful stabilizing influence. The Moon gravitational influence allows Earth's obliquity to oscillate by about 1.3 degrees about its mean position of 23.3 degrees. If the Moon did not exist, Earth's obliquity would evolve chaotically, similarly to what happened on Mars. Chaotic evolution of obliquity would have had dire consequence for the climate of the extrasolar planet. Hence the importance of the presence of a moon around a planet whose habitability had been determined by its distance from his star, and the presence of an atmosphere and its composition.

7.2.1 Methods to detect an extrasolar planet capable of sustaining life

Astronomers detected an extrasolar planet by measuring its gravitational influence on the parent star's motion. They used the Doppler effect to analyze motions and properties of the star and the planet. Both are orbiting a common center mass. They gravitationally attract each another causing the star to wiggle around the point of common center of mass. As the star approaches the observer, the light emitted shifts to higher frequencies (blue shift). As it recedes, light shifts to lower frequencies (red shift). The Doppler technique gives information on the star's velocity toward or away from the planet (star radial velocity). The planet's velocity can then be derived from the star's radial velocity. Once the star's and the planet's velocities and the star's mass are known, the planet's mass can be calculated. This method has the advantage of being applicable to stars with a wide range of characteristics. One disadvantage is that it cannot figure a planet's true mass. It can only set a lower limit on the planet mass because the angle of the orbital plane is not known. Another disadvantage is that small, Earth-like, extra-solar planets are harder to detect than giant massive planets. About nineteen percent of exoplanets were discovered using this method.

A second method (transit method) consists of observing variations in the star's apparent luminosity as a planet crosses it. The variation in star luminosity is equal to the ratio of planetary and stellar disk areas. The planet's size is found from the star luminosity variation and size. The star's orbital size and temperature enable to calculate the planet's characteristic temperature[11], which is a gauge of its habitability. This method makes possible the study of the planet's atmosphere and its composition[12]. Seventy eight per cent of the discoveries were made using this method. The transit method has two major disadvantages. First, planetary transits are only observable for planets whose orbits are aligned with Earth, but less than one per cent of stars have this desired orbit. If the angle is different, the planet never appears to pass in front of its star. Second, this method suffers from a high rate of false detection. A 2012 study[13] found that the rate of false detection for transits, observed by the Kepler mission, was as high as forty per cent in a single planetary system.

A third method is called transit-time variation. It is a method for detecting extrasolar planets by observing variations in the timing of a planet's transit. Transit time variation enables to detect more planets even if these planets do not transit. When multiple planets are present, each one perturbs the others' orbits. Small variations in transit time of a transiting planet show another planet's presence. This method is useful in planetary systems far from the Sun, where radial velocity methods cannot detect them due to the low signal-to-noise ratio. The main drawback of the transit timing method is that usually not much can be learned about the planet itself. Transit timing variation can help to determine the maximum mass of a planet. In most cases, it can confirm if an object has a planetary mass, but it does not put narrow constraints on its mass.

---

[11]Secondary eclipse enables the measurement of the planet's radiation. If the star's photometric intensity during secondary eclipse is subtracted from its intensity before or after, only the signal caused by the planet remains. It is then possible to measure the planet's temperature

[12]When a planet transits the star, the star's light passes through the upper atmosphere of the planet. Planetary atmosphere can be detected by measuring the starlight's polarization as it passes through or is reflected off the planet's atmosphere. By studying high-resolution stellar spectra, one can detect elements present in the planet's atmosphere.

[13]See Santerne A. et al. (2012).

A fourth method is the gravitational microlensing. It exploits one of the predictions Einstein theory of general relativity; i.e., light follows the curvature of space-time. The star mass deflects light because it warps space-time. Gravitational microlensing occurs when a star's gravitational field acts as a lens deflecting the light of a distant background star. If the foreground star has a planet, then that planet's own gravitational field causes an asymmetrical perturbation in the signal. This effect occurs when the two stars are aligned. Since microlensing observations do not rely on radiation received from the lens object, this effect enables the study of massive objects no matter how faint. But lensing events are brief, lasting for weeks or days, as the two stars and Earth are moving relative to each other. Since this requires an improbable alignment, many distant stars must be continuously monitored to detect extrasolar planets. This method is most fruitful for planets between Earth and the center of the galaxy as the galactic center provides many background stars. Only one and a half per cent of the discoveries were made using this method.

A fifth method under study would allow the detection of a magnetosphere of an extrasolar planet. These studies cover both indirect and direct methods to detect a magnetosphere. An indirect method infers the presence of a magnetosphere from the way hydrogen is escaping from the extrasolar planet. In 2014, a magnetic field around HD 209458 b[14] was inferred with this method. The magnetic field was estimated to be about one tenth as strong as Jupiter's. A direct method is basically built on the premise that planets in the solar system, including Earth, generate electromagnetic waves as space plasma interacts with their magnetosphere. In theory, with the right equipment, it should be possible to detect an extrasolar planet's magnetic field. But terrestrial extrasolar planet magnetic field strength estimates predict signals too weak and with too low frequency for radio telescopes in operation today to detect. But, timely coronal mass ejections may boost signal strengths enough, and a radio telescope array placed on the Moon could detect the low frequencies emitted.

---

[14]HD 209458 b is an extrasolar planet that orbits the solar analogue HD 209458 in the constellation Pegasus, some 150 light-years from the Solar System. The radius of the planet's orbit is seven million kilometers, about 0.047 astronomical units, or one eighth the radius of Mercury's orbit. This small radius results in a year that is 3.5 Earth days long and an estimated surface temperature of about 1,000 °C. Its mass is 220 times that of Earth (0.69 Jupiter masses) and its volume is some 2.5 times greater than that of Jupiter. The high mass and volume of HD 209458 b indicate that it is a gas giant.

7.2.2 Where antimatter propulsion will bring humanity

Figure 7.1 shows the star systems within 10/15 light years from earth. There are many criteria by which to choose the prime targets for an interstellar probe mission. One criterion would be to choose as a target a star similar to our Sun. Since our Sun is positively known to possess a planetary system, it can be assumed that stars similar to the Sun (same spectral type, mass, radius and luminosity) also have some possibility of possessing a planetary system. The argument advanced here is that, by using a star's similarity to the Sun as a selection criterion, we would be choosing target stellar systems not only for their potential planetary systems but also on their ability to support life.

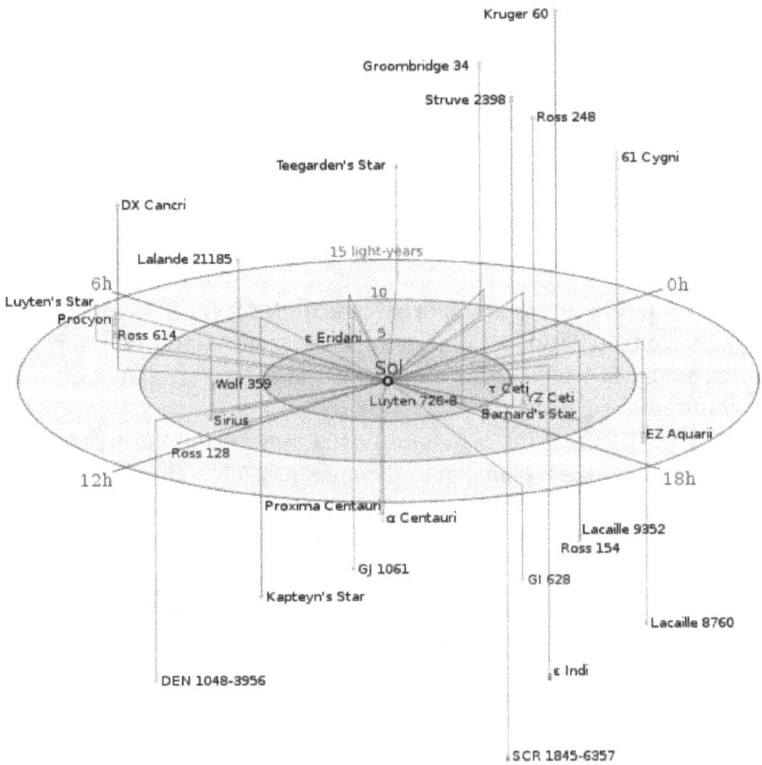

Figure 7.1 – Closest Star Systems to the solar system
Credit:
https://en.wikipedia.org/wiki/List_of_nearest_stars_and_brown_dwarfs#/media/File:Nearby Stars_(14ly_Radius).svg

Another criterion is to look at stars different from our Sun. A class of stars, M-class dwarf stars, has attracted the attention of astronomers. These stars are much smaller than the Sun and so thrifty with their fuel resources that can last trillion of years. They are more numerous than heavier Sun-like stars. Hence these stars could be ideal in the search of habitable planets. But they present a problem, which is still debated by astronomers. Since their habitable zone is five times narrower than the one around the Sun, it is uncertain that planets within this zone could maintain stable orbits over long periods of time. If a planet orbit is too close to a star, it could become tidally locked after few billion years from its formation, thus presenting the same face inward that could make it not hospitable for life. However, some astrobiologists maintain that it is possible for liquid water to exist in the twilight zone between the face permanently exposed to its sun and the dark face. The verdict on this issue is still open.

As of September 2017, thirteen stars (Alpha Centauri A and B, 82 Eridani, Epsilon Eridani, Gliese 876/Ross 780, CD-44 11909/Gl 682, CD-46 11540/Gl 674, CD-49 13515/Gl 832, Gl 229, Groombridge 34 A, Kapteyn's Star 229, Lalande 21185, Tau Ceti, and Wolf 1055 B) may possibly host Earth-type planets in stable orbits within their respective circumstellar habitable zones.

Alpha Centauri is the closest star system to the Solar System, being 4.37 light-years from the Sun. It consists of three stars: Alpha Centauri A and Alpha Centauri B, which form the binary star Alpha Centauri AB, and a small and faint red dwarf, Alpha Centauri C or Proxima Centauri. Alpha Centauri A has 1.1 times the mass and 1.519 times the luminosity of the Sun, while Alpha Centauri B is smaller and cooler, at 0.907 times the Sun's mass and 0.445 times its visual luminosity. Proxima Centauri is a red dwarf of spectral class M6 with an absolute magnitude of +15.60, which is only a small fraction of the Sun's luminosity. By mass, Proxima Centauri is calculated at about one-eighth that of the Sun. It is about 13,000 astronomical units[15] (AU) away from Alpha Centauri AB. A paper by Kervella, et al. (2017) showed that, based on high precision radial velocity measurements and with a high degree of confidence, Proxima and Alpha Centauri are in fact gravitationally bound.

[15]The astronomical unit (AU) is defined by the International Astronomical Union as exactly 149,597,870 kilometers. It is approximately the average distance between the Earth and the Sun

Table 7.1 – Possible targets for interstellar missions

Stellar System	*Distance from Earth in light years*	
Alpha Centauri	4.3	Triple system (G0, K5, M5). Component almost identical to Sun. High probability of planets in habitable zone.
Barnard's Star	6	Very small, low luminosity red dwarf (M5). Closest system known to have one, and perhaps two or more planetary companions.
Epsilon Eridani	10.8	Single star system; slightly smaller and cooler than the Sun (K2), may have a planetary system similar to the solar system.
Ross 128 b	11	Red dwarf star, which seems to be an inactive. Even though Ross 128 b is twenty times closer to its star than Earth is to the sun (the orbit is just 9.9 days), chances are that Ross 128 b is temperate, particularly when compared with Proxima b.
Tau Ceti	11.8	Single star system, similar in size and luminosity to the Sun (G4). High probability of possessing a "Solar-like" planetary system.
Kapteyn	12.6	Red dwarf star (M2) known to have a planet in the habitable zone, Kapteynb. One other planet has been detected.
Wolf 1061	13.8	M class red dwarf star. It is known to have three planets orbiting this star. One of the planet is located near the inner edge of the star's habitable zone.
Gliese 832	16.1	Red dwarf star with confirmed planetary system composed of four planets.

Credit: author

On October 16, 2012, a team of astronomers[16] announced the discovery of a planet with around 1.13 +/- 0.09 Earth-masses in a very hot and tight, circular orbit around Alpha Centauri B. Due to the close binary orbital interactions of the host star with Alpha Centauri A, the astronomers were only able to detect the radial-velocity variations of host star B that were caused by the 3.236-day orbit of the planet (with a semi-major axis of 0.04 AU) only after more than four and a half years of careful observation. Given that the presence of one close-orbiting planet usually indicates the presence of others, many astronomers are now expected to devote more resources to detecting such potential planets around Alpha Centauri B.

In August 2016, the European Southern Observatory announced the discovery of an extrasolar planet orbiting Proxima Centauri within the habitable zone. The planet was found using the radial velocity method, where periodic Doppler shifts of spectral lines of the host star suggest an orbiting object. The planet orbits the star at a distance of roughly 0.05 AU with an orbital period of approximately 11.2 Earth days. It has an estimated mass of at least 1.3 times that of the Earth. Its habitability has not been established, though it is unlikely to be habitable since the planet is subject to stellar wind pressures of more than 2,000 times those experienced by Earth. This radiation and the stellar winds would likely blow any atmosphere away, leaving the under surface as the only potentially habitable location on that planet. Moreover, it is close enough to its host star that it might be tidally locked[17]. In this case, it is expected that any habitable areas would be confined to the border region between the two extreme sides, generally referred to as the terminator line, since it is only there that temperatures might be suitable for liquid water to exist. A recent paper[18] indicates that a ring of cold dust circles the star. It suggests that Proxima Centauri may have a multiple planet system with a rich history of interactions that resulted in the formation of a dust belt.

Barnard's Star is a low-mass red dwarf around six light-years from Earth. At seven to twelve billion years of age, Barnard's Star is older than the Sun. It might be among the oldest stars in the Milky Way galaxy. In 1969, Peter van de

---

[16]See Dumusque X. et al. (2012)

[17]A tidally locked body in synchronous rotation takes just as long to rotate around its own axis as it does to revolve around its partner. This state can result from the gravitational gradient (tidal force) between two co-orbiting bodies, acting over a sufficiently long period of time. In the case where the orbital eccentricity is exactly zero, tidal locking results in one hemisphere of the revolving object constantly facing its partner, an effect known as synchronous rotation.

[18]See Guillem Anglada et al. (2017).

Kamp[19] claimed detection of one or more planets with masses comparable to Jupiter. For a decade, astronomers accepted this claim, but two important papers in 1973 contested that claim. Null results for planetary companions continued throughout the 1980s and 1990s. Observations based on interferometry from the Hubble Space Telescope in 1999 were the latest. While research has restricted the possibility of the existence of planets around Barnard's Star, it has not ruled them out.

Epsilon Eridani is the third-closest individual star or star system to the Sun. It is situated a distance of 10.5 light-years from the solar system. It has an apparent magnitude of 3.73. The star is estimated to be less than a billion years old. Because of its relative youth, Epsilon Eridani has a higher level of magnetic activity than the present-day Sun, with a stellar wind thirty times as strong. Observations from NASA's Spitzer Space Telescope suggest that the star actually has two asteroid belts and a cloud of exozodiacal dust. One belt sits at approximately the same position as the one in the Solar System, orbiting at a distance of about 3.00 AU from Epsilon Eridani. The second, denser belt, most likely also populated by asteroids, lies between the first belt and the outer comet disk. The structure of the belts and the dust disk suggests that more than two planets in the Epsilon Eridani system are needed to maintain this configuration[20]. The habitable zone has been estimated between 0.47 and 0.91 AU. Its inner edge could be around 0.507 AU from the star, while the outer edge lies at around 1.005 AU.

Epsilon Eridani has been the target of many attempts to search for planetary companions. Its chromospheric activity and variability mean that finding planets with the radial velocity method is difficult, because the stellar activity may create signals that mimic the presence of planets. Attempts at direct imaging of potential exoplanets have proven controversial to date. The discovery of Epsilon Eridani b was announced in 2000, but has remained controversial. A comprehensive study[21] in 2008 called the detection tentative and described the proposed planet as long suspected but still unconfirmed. The true mass of this planet remains unknown, but it can be estimated based on the displacement effect of the planet's gravity on its parent star[22]. Estimates for the value of its mass range from 0.60 Jupiter masses to 1.06 Jupiter masses.

---

[19]He had detected a perturbation in Barnard's Star's proper motion consistent with having one or more planets comparable in mass to Jupiter. See Van de Kamp P. (1969).

[20]See D. Backman et al. (2008).

[21]See Backman, D.; et al. (2000)

[22]The component of the displacement along the line of sight to Earth is known, which yields a value of the mass from the formula $m \sin i$, where m is the mass of the planet and i is the orbital inclination.

This planet is estimated to have a relatively circular orbit that keeps it just outside the inner asteroid belt at an average distance of 3.4 AU from its star, well outside the habitable zone.

An international team of astronomers published a paper detailing the planet's discovery in the journal Astronomy and Astrophysics. Ross 128 b is a little farther away than Proxima b, the closest planet beyond our solar system, but it probably has better odds of being inhabitable. Ross 128 b orbits the star Ross 128 just 11 light-years away. Unlike other close rocky planets, such as those orbiting Trappist-1 and Proxima Centauri, Ross 128 b is set apart because its host star does not regularly blast it with potentially sterilizing flares of radiation. Ross 128 b's host star is an M dwarf. The problem with M dwarfs is that many, such as Trappist-1 and Proxima Centauri, have a nasty habit of sending out big flares that could erode the atmospheres of orbiting planets. This bathes those surfaces in radiation that makes it difficult for life as we know it to emerge or survive. Ross 128, on the other hand, is what's known as a "quiet" M dwarf, meaning it does not seem to be shooting off the killer flares seen from other stars like it. The research team estimates that Ross 128 b receives just 38 percent more radiation than Earth does. Still, while Ross 128 b is a fascinating discovery, it's way too early to say anything definitive about its potential. The new findings have plenty of shortcomings of their own. For one, the European Southern Observatory team has no idea yet whether the new exoplanet sits in Ross 128's habitable zone. As with every other exoplanet we like to call "potentially habitable," there is no indication whether the planet has water on its surface. There is not even any sign as to whether the planet possesses other elements essential to life as we know it, like atmospheric oxygen. Some of these questions could be resolved relatively soon. The ESO's Extremely Large Telescope could find out whether Ross 128 b possesses oxygen or other chemical markers that could be signs of habitability or extant life.

Tau Ceti is located about 11.9 light-years away from the solar system. It is a main sequence, yellow-orange dwarf. Smaller and dimmer than the Sun, the star has around 81 to 92 percent of Sun's mass. The luminosity of Tau Ceti is equal to only 55% of the Sun's luminosity. A terrestrial planet would need to orbit this star at a distance of about 0.7 AU in order to match the solar-insulation level of Earth. This is approximately the same as the average distance between Venus and the Sun. The habitable zone for this star is within a radius of 0.55–1.16 AU The lower abundance of iron[23] indicates that Tau Ceti is almost certainly older than the Sun. Its age had previously been estimated

[23]The chemical composition of a star provides important clues to its evolutionary history, including the age at which it formed. Younger stars will tend to have a higher portion of heavy elements in their atmospheres than do the older stars.

to be about 10 Ga[24] but is now thought to be around half that at 5.8 Ga, comparing with 4.57 Ga for the Sun.

Tau Ceti has been a target of a radial velocity planetary searches. On December 19, 2012, evidence[25] was presented that suggests a system of five planets orbiting Tau Ceti. The planets' estimated minimum masses are between two and six times the mass of Earth and their orbital periods range from 14 to 640 days. One of them, tentatively named Tau Ceti e, appears to orbit about half as far from Tau Ceti as Earth does from the Sun. With Tau Ceti luminosity of 52% that of the Sun and a distance of the planet from the star of 0.552 AU, the planet would receive 1.71 times as much stellar radiation as Earth does, slightly less than Venus with 1.91 times Earth's. Nevertheless, some research[26] places it within the star's habitable zone. The planet has an orbital period of 168 days and has a minimum mass of 4.3 Earth masses. Because the minimum mass of a super-Earth is 5 Earth masses, Tau Ceti e may be Earth sized. But a team of astronomers[27] have detected a large and relatively dense, cold dust disk around Tau Ceti . This disk may be a ring that extends mostly between ten and around fifty five AU from the star. Tau Ceti dust disk may have around ten times more cometary and asteroidal material than is currently found in the Solar System, likely experiencing relatively frequent bombardment from asteroids and comets. As a result, some astronomers have speculated that it is likely that with so many large impacts, large and complex forms of Earth-type multi-cellular life may not have had the opportunity to evolve and persist on inner terrestrial planets orbiting this star.

Kapteyn's Star is a class M1 red dwarf about 12.76 light years from the solar system. The star has around 7.4 percent of the Sun's abundance of iron and a weighted mean metallicity abundance of elements heavier than hydrogen and helium of almost 10.5 percent of the Sun. Due to Kapteyn's proximity to the Sun and its deduced ancient origin from outside the galactic disk, the system has been an object of high interest among astronomers. It may have around 29 to 39 percent of the Sun mass, around a third of its diameter, and less than 4/1,000th of its luminosity. According to a 2014 analysis[28], the inner edge of Kapteyn's habitable zone is located relatively close from the star at around 0.126 AU from the star, while the outer edge lies farther out at around 0.236 AU.

---

[24]Gigaannus: a unit of time equal to 10^9 years.
[25]See M. Tuomi (2012).
[26]See A. M. Torres (2012).
[27]See Greaves et al. (2004)
[28]See Anglada-Escudé et al. (2014).

On June 3, 2014, a team of astronomers[29] reported the discovery of two super-Earths orbiting this ancient star. Kapteyn b has at least 4.8 Earth mass. It might support liquid water on its surface, completing its orbit within only 48.6 days at an average orbital distance of 0.17, with an orbital eccentricity of 0.21. The planet is around 11.5 billion years old. This makes it two and a half times older than Earth and just two billion years younger than the universe itself. Kapteyn c is even more massive at a minimum of 7.0 Earth mass and its year lasts 121.5 days at an average orbital distance of 0.31, with an orbital eccentricity of 0.23, and should be too cold to support liquid water. Both planets are in a dynamical state called apsidal co-rotation, which implies that the planets are dynamically stable over long time-scales.

Wolf 1061 is an M class red dwarf star located about 13.8 light years away from the solar system. Wolf 1061 is very stable and most likely does not have any significant activity such as sunspots or flares. A seven years study found no evidence of photometric transits and confirms the radial velocity signals are not due to stellar activity. The Habitable Zone estimate for the system lies between 0.11-0.21 AU and 0.09–0.23 AU. n December 2015, a team of astronomers announced the discovery of three planets orbiting Wolf 1061[30]. The planets (Wolf 1061 b, c,d) were detected by analyzing ten years of observations of the Wolf 1061 system by the HARPS spectrograph at La Silla Observatory in Chile. The team used archive radial velocity measurements of the star's spectrum in the HARPS data and, along with eight years of photometry from the All Sky Automated Survey, discovered two definite planets with orbital periods of around 4.9 and 17.9 days and a very likely third with a period of 67.3 days. All three planets have masses low enough that they are likely to be rocky planets similar to the inner planets of the Solar System although their actual sizes and densities are currently unknown. This information could be determined if the planets happen to transit in front of Wolf 1061 when viewed from Earth. Because all three planets orbit close to the star and have short orbital periods, there is a chance that this will occur[31]. One of the planets, Wolf 1061 c, is located near the inner edge of the star's habitable zone. It is one of the closest known potentially habitable planets to Earth after Proxima b. Wolf 1061 d, could be marginally habitable depending on its atmosphere's composition as it orbits just beyond the habitable zone.

---

[29]See Anglada-Escudé et al. (2014)

[30]See www.Newt.phys.unsw.edu.au.

[31]The same team of astronomers estimated the chances of a transit at around 14% for planet b, 6% for planet c, and 3% for planet d.

Gliese 832 is a red dwarf star at a distance of 16.1 light years away from the solar system. This star has been estimated[32] to have around forty five percent of Sun mass, 43 to 48 percent of its diameter, and about 0.7 percent of its visual luminosity and 2.8 percent of its bolometric luminosity. It is chromospherically quiescent and appears to have only around 49 percent of Sun abundance of iron relative to hydrogen. The star is roughly 9.5 billion years old. In order to be warmed sufficiently have liquid water at the surface, an Earth-type rocky planet would have to be located very close to such a red dwarf star as cool and dim as Gliese 832.

Gliese 832 hosts two known planets. In September 2008, it was announced[33] that a Jupiter-like planet, now designated as Gliese 832 b, had been detected in a long-period, near-circular orbit around this star. This planetary companion has at least 64 percent of Jupiter's mass. It lies at an average distance of 3.4 AU. Planet b orbit is completed within 9.4 years. In 2014, a second planet was discovered by astronomers at the University of New South Wales. This one is believed to be of super-Earth mass and has since been given the scientific name Gliese 832 c. Planet c has about 5.4 Earth-mass, an orbital period of 35.68 days at a semi-major axis of 0.163 AU, and an orbital distance near the inner edge of its host star's habitable zone. If it has an Earth-like terrestrial atmosphere, planet c may have Earth-like temperatures with possibly with large seasonal shifts. But a denser atmosphere as would be expected for massive super-Earths could easily make the planet too hot for life and more akin to a super-Venus instead.

This is what is known up to now. Detecting more planets will become easier in the future. The European Southern Observatory's next-generation spectrometer ESPRESSO, which looks for Earth-like planets, will come online in 2017. It will offer radial-velocity measurements several times more precise than those used up to now[34]. Soon, both terrestrial and space-based telescopes will provide analyses of the spectrum of planets' light to identify chemical elements, such as oxygen, in their atmospheres. The Extremely Large Telescope (ELT) is the world's largest optical/near-infrared extremely

---

[32]See NASA Star and Exoplanet Database.

[33]See J. Bailey et al. (2008).

[34]The ELT will use a novel design with a total of five mirrors. The first three mirrors are curved (non-spherical), and form a three mirror anastigmat design for excellent image quality over the 10 arcminute field of view (one third of the width of the full Moon). The fourth and fifth mirrors are (almost) flat, and provide adaptive optics correction for atmospheric distortions (mirror 4), and tip-tilt correction for image stabilization (mirror 5). The fourth and fifth mirrors also send the light sideways to one of the focal stations at either side of the telescope structure, allowing multiple large instruments to be simultaneously mounted.

large telescope. It is now under construction and should be operational in 2024 . The ELT will search for extrasolar planets. This will include not only the discovery of planets down to Earth-like masses through indirect measurements of the wobbling motion of stars perturbed by the planets that orbit them, but also the direct imaging of larger planets and possibly even the characterization of their atmospheres. The telescope will attempt to image Earth like exoplanets.

A number of space telescopes will be launched in the next few years. The James Webb Space Telescope (sometimes called JWST) will be launched in 2018. It will be a large infrared telescope with a 6.5-meter primary mirror. Among the JWST's primary scientific missions are the study planetary systems and the origins of life. The Transiting Exoplanet Survey Satellite (TESS) is a planned space telescope for NASA's Explorers program, designed to search for exoplanets using the transit method. It is planned for launch in March 2018. TESS will discover thousands of exoplanets in orbit around the brightest stars in the sky. In a two-year survey of the solar neighborhood, TESS will monitor more than 200,000 stars for temporary drops in brightness caused by planetary transits. This first-ever space borne all-sky transit survey will identify planets ranging from Earth-sized to gas giants, around a wide range of stellar types and orbital distances.

The Square Kilometer Array (SKA) is a large multi radio telescope project to be built in Australia, New Zealand, and South Africa. Once built, it would have a total collecting area of approximately one square kilometer. It will operate over a wide range of frequencies, and its size will make it fifty times more sensitive than any other radio instrument. It will be able to survey the sky more than ten thousand times faster than ever before. Construction of the SKA is scheduled to begin in 2018 for initial observations by 2020. The SKA, global project with twelve member countries, aims to answer fundamental questions about the origin and evolution of the Universe. But the SKA will be capable of detecting extremely weak radio emissions, which may help in detecting magnetosphere around extrasolar planets.

7.3 The exploratory and colonizing missions

In the next twenty to thirty years, astronomers may find many other extrasolar planets. Among them, real targets for interstellar missions, i.e. planets within five to fifteen light years from Earth, in habitable zone, with Earth-like masses, oxygen in their atmospheres, and a magnetosphere. This discovery will be the crucial driver for interstellar exploration and colonization. It will provide the incentives to invest in technologies to produce antimatter at reasonable costs. But, before sending human missions, robotic missions should be organized.

NASA is in the earliest stages of planning an exoplanet expedition in 2069, set to mark the 100th anniversary of its first crewed Moon landing, Apollo 11[35]. The basic concept for the 2069 exoplanet expedition was presented by JPL's Anthony Freeman at the 2017 American Geophysical Union conference. He later described its current status as "nebulous" in an interview with New Scientist[36]. At present, there is some debate as to whether Alpha Centauri or another system will be selected, as there are several candidates being considered. It will also be no small feat to develop the propulsion technology necessary to get this project off the ground. Antimatter-matter collisions, nuclear propulsion, and laser-powered light sails are all being considered as technologies that could potentially help a craft reach the speeds that are required. The practicalities of carrying out a mission that will last decades upon decades offer up their own challenges. There are various technologies that need to be put in place, including optical communication, re configurable spacecraft, compact space-qualified 3D printing capabilities, genomic programming, and artificial intelligence.

The goals of these robotic missions are not only to proof alien life, but also to allow for a greater knowledge of the exoplanet in question, opportunities to study matter and radiation encountered en-route, and tests of the theory of relativity. Robotic missions will find if life forms, including intelligent life, exist on a planet. We believe that intelligent life does not exist on extrasolar planets near Earth, since the search for intelligent life's signals has so far yielded no result[37]. But no one knows what lies around the corner. We do not want to repeat the mistakes of Earth's colonizers who destroyed entire civilizations. After robotic missions, human missions for a planet's colonization would be ready to start.

---

[35]A small team based at the Jet Propulsion Laboratory (JPL) hopes to send a spacecraft to a distant planet in search of life. In 2016, a funding bill was passed that called upon NASA to Investigate methods of interstellar travel that could reach at least 10 percent of the speed of light by 2069. It also requested a mission to Alpha Centauri, the closest star system to our own. At present, there's some debate as to whether Alpha Centauri or another system will be selected, as there are several candidates being considered.

[36]New Scientist (December 2017).

[37]First attempts to search intelligent life began more than half a century ago. Francis Drake started project Ozma and continued with the more ambitious and sophisticated project SETI (Search for Extraterrestrial Intelligence). SETI searches for intelligent life with radio telescopes to detect their signals. These projects analysed signals from many stars near the solar system. But there has been absolute silence.

7.3.1 Motivations for human missions

Some observers contend that the main reasons behind manned interstellar exploration are the survival of the human species and the spreading of life in the universe. Physicist Stephen Hawking once said:

> "The long-term survival of the human race is at risk as long as it is confined to a single planet. Sooner or later, disasters such as an asteroid collision or nuclear war could wipe us all out. But once we spread out into space and establish independent colonies, our future should be safe. As there is not anywhere likes the Earth in the solar system, so we would have to go to another star."

We argue that the survival of the human species is a very long term threat and, as such, may not be the main driver for interstellar exploration. Historically, voyages of exploration and discovery and migrations have characterized and distinguished humanity since the dawn of its history. Humanity is a wandering race. Throughout the history of Homo sapiens motivations for its wandering have been various. During the early history of our species, migrations out of Africa were most likely the result of the extremely variable climate in the last part of Pleistocene rather than conscious motivations. People moved at a very slow pace following the animal herds that moved because of changing climate. Home sapiens appear to have occupied all of Africa about 150,000 years ago, moved out of Africa 70,000 years ago, and spread across Australia, Asia and Europe by 40,000 years BC.

Development of agriculture and domestication of animals changed the pace of migration. The food surplus generated in these societies allowed the sustaining of groups of people who were not engaged in the production of food. One group of people, craft specialists, generated the technologies allowing faster traveling. The Phoenicians explored all the Mediterranean Sea in search for trade and, in the process, established many colonies. They ventured out north to the Atlantic Ocean toward Ireland and England, and they circumnavigated Africa. Claims of pre-Columbian trans-oceanic contact between indigenous peoples of the Americas and peoples of other continents before the arrival of Christopher Columbus in 1492 are numerous but sometimes dubious. Only one instance of pre-Columbian European contact is established beyond reasonable doubt. Viking journeys to North America are supported by both historical and archaeological evidence[38].

[38]See Helge and Anne Ingstad (2001).

Modern historians[39] of the Polynesian people before European contact estimate that their journey stretched from 3000 BC to around 800 AD. Around 3000 BC, speakers of Austronesian languages, probably on Taiwan, mastered the art of long-distance canoe travel. They spread south to the Philippines and Indonesia, and east to Micronesia and Melanesia. They then branched off and occupied Polynesia to the east. Dates and routes are uncertain. They started most likely from the Bismarck Archipelago and reached Samoa and Tonga around 1500 BC. By 1000 AD they were in the Marquess Islands and by 300-800 AD in Tahiti. They arrived at Easter Island, their easternmost point, around 300-800 AD. The same date range is estimated for their arrival in Hawaii, far to the north and distant from these other islands.

It is evident that these voyages were not made by chance. Transfer of many species of crops and livestock all over the Pacific islands prove that the settlements were done by colonists carrying products and animals from their homeland, deemed essential for the survival of the new colonies. There is still a debate on the motivations for these migrations. Several arguments have been put forward for explaining these migrations. One refers to population growth in the new settlement. As population grew in the new settlement, it may have put pressures on the limited agricultural resources of the island, inducing a group of colonists to move either into the unknown or to new island previously discovered by voyagers sailing upwind on a predetermined bearing. Another argument is that human activities, as deforestation in the new settlement, posed pressure on a fragile ecosystem thus creating a crisis that could only be solved by migration. In reality, it was probably a combination of these factors that was behind the spreading of the Polynesian people throughout the Pacific Ocean.

European voyages of exploration started in the early 15th century with the discovery of America and culminated in the 17th century[40]. Though the desire to simply explore the unknown and discover new knowledge is a typical human trait, the world's famous explorers often lacked the funding needed for a ship, supplies, and a crew to get underway on their journeys. As a result, many turned to their respective governments, which had their own desires for the exploration of new areas. Many nations were looking for goods such as silver and gold but one of the biggest reasons for exploration was the desire to find a new route for the spice and silk trades. When the Ottoman Empire took control of Constantinople in 1453, it blocked European access to the area, severely limiting trade. Although Columbus did not find a direct western route to China, he discovered America.

---

[39]See Kirck P.V. (2000).

[40] See Cowie L. (1985)

Voyages of exploration did not cease entirely in the 17th century. Eastern Australia was not discovered until 1770 and the Arctic and Antarctic areas were not heavily explored until the 19th century. Much of Africa also was also unexplored until the 19th and even early 20th centuries. Some of these voyages were motivated by the acquisition of knowledge, as James Cook scientifically motivated trip to Tahiti and Darwin's voyage around the world aboard the Beagle. Other voyages of exploration were motivated by the acquisition of land for future colonization. In 1770, James Cook charted the East Coast of Australia for Britain and returned with accounts favoring colonization at Botany Bay (now in Sydney) and New South Wales. Similarly, voyages of exploration in Africa in the 19th century were carried out by explorers, often motivated by religious purposes, knowledge and sense of adventure, but financed by governments interested in the colonization of the African continent because of its riches in natural resources.

Voyages of exploration created the infrastructure allowing large migration and settlements of the discovered lands. First, Spain and Portugal colonized Central and South America, and part of North America, notably California, Florida and Texas. Inspired by the Spanish riches from colonies, other European countries[41] established permanent settlements in North America and the Caribbean islands. Initially, a large portion of the immigrants were indentured servants looking for new life in the overseas colonies. As American landowners were in need of laborers and were willing to pay for a laborer's passage to America if they served them for several years. By selling passage for five to seven years worth of work, laborers could hope to start out on their own in America.

Other groups of colonists came to the Americas searching for the right to practice their religion without persecution. The third large group of immigrants was forced immigrants or slaves. Slave trade started in Spanish and Portuguese colonies. As the native populations declined, they were often replaced by Africans imported through a large commercial slave trade. It was followed by slave trade to North American colonies. The total slave trade to islands in the Caribbean, Brazil, and Mexico and to the United States is estimated to have involved 12 million Africans.

The brief analysis of humanity wandering indicates that motivation for voyages of discovery and migration can be divided into two types: internal motivation and external motivation. Internal motivation refers to motivations that are driven by an interest or enjoyment in the task itself, and exists within the individual rather than relying on external pressures. External motivation

---

[41]Kenneth A. (1984).

refers to the performance of an activity in order to attain an outcome, whether or not that activity is also internally motivated. Hereafter this section will analyze the internal and external motivations behind interstellar exploration.

Internal motivations

Internal motivation, as the name suggests, comes from within. An internally motivated person is inspired by staying true to their own core values and doing the things which they love the most. Today, as yesterday, personal values are largely diversified among the human population. There are many individuals who place high value in individual freedom, personal control and self-realization, and in living free from the constraints imposed by the conditions of overcrowded conglomerates: the modern version of the frontier men and women. A recent survey[42] indicates that these people represent about twenty five per cent of the population in advanced industrial countries.

Recent research has identified a DRD4 gene's variant that controls dopamine. The variant gene is part of the dopamine system. It facilitates transmission of signals among neurons and plays a major role in the brain network responsible for attention and reward-driven learning. The DRD4-7R allele blunts dopamine signalling, which enhances individuals' reactivity to their environment. It drives the search for the new and risk taking. This variant evolved in the distant past enhancing the survival prospects of those who possessed it. Someone[43] has suggested that resource depletion or changing environments might have selected for individuals with this variant. Such adaptation most likely played a role in the out-of-Africa event. Later on, local cultural environments influenced allele frequencies associated with such behavior. Preliminary results of this research show this allele present in twenty per cent of humanity, and this number may eventually increase due to ongoing research.

Others place high values on the search of knowledge and new opportunities, or on the simple desire for adventure and exploration, the modern version of Dante's Ulysses. The urge to know what is over the next hill is one of the most common feelings humanity share, whatever their backgrounds. Self-determination theory suggests that people are motivated to grow and change by innate psychological needs. The theory identifies three key psychological needs that are believed to be both innate and universal: the need for competence, contentedness, and autonomy. How exactly do people go about fulfilling these three needs? It will stimulate the interest in science of the best and the brightest among young people because of the challenges of developing

[42]See Klinenberg E. (2012).
[43]See G. Gugliotta (2008).

the technologies for interstellar exploration. The fact that one does not know yet what is out there and that space represents our final limitation is reason enough to strive to master it.

It is important to realize that the psychological growth described by self-determination theory does not simply happen automatically. While some people might be oriented toward such growth, it requires continual sustenance. In analyzing the inner motivations of interstellar exploration and colonization, it is thus important to review the social impacts of human space exploration. The social impacts of space exploration have been extensively analysed[44]. There is yet no agreement on social and cultural effects of spaceflight. But there is a general consensus on space flight's effects on the evolving construction of our worldview. Historically, space exploration and technology helped in the development of human consciousness with respect to global issues and the need to solve them by extensive international cooperation. The discovery of an Earth-like planet and the challenges posed by interstellar exploration would help in identifying ourselves as inhabitants of a single planet. In turn, this will facilitate overcoming sectarianism, fundamentalism and racism paving the way to a planetary culture. This does not mean that local cultures and customs would not continue to thrive. It only means that a planetary culture would be superimposed on this mosaic of cultures and in the process enrich them.

External motivations

External motivations are related to the possibility that Earth will no longer define the limit of growth. It may be simple economics that ultimately provides the impetus for sustained effort at living in extraterrestrial environments. Space is literally the largest store of natural resources in the universe, and it remains completely untapped. Resources in space, both in materials and energy, are enormous. The Solar System alone has, according to different estimates, enough material and energy to support anywhere from several thousand to over a billion times that of the current Earth-based human population. Outside the Solar System, other stars in the Milky Way provide opportunities for both colonization and resource collection.

Another argument for extrasolar planets colonization is to mitigate the negative effects of overpopulation on Earth. In the solar system there are no planets similar to Earth. Even Mars will not offer the conditions for a large movement of humans, but only the possibility of building a permanent base for a small number of people. If an extrasolar planet capable of sustaining life were found, it could accommodate in time larger scale migrations from Earth

[44]See NASA History Division (2007).

and the use of its resources. Large profit opportunities would arise through new forms of space commercialization. It would reduce the stress on Earth environment due to overpopulation.

The first wave of colonists will face the challenge of surviving until the spaceship's return from Earth. They must have the same psychological and professional profiles as those selected for Mars colonization. The settlers' number depends on how many years the colony will stay isolated before the spaceship's return to Earth. Researchers in conservation biology adopt "50/500" rule of thumb. To prevent unacceptable rates of inbreeding, this rule requires a short-term population size of fifty. The same rule needs a long-term population size of five hundred to maintain overall genetic variability[45]. We thus assume that the number of colonists should be between fifty to one hundred. This enables normal reproduction for around 30-60 generations.

What will settlers bring with them? It depends on the particular planet's characteristics as identified by robotic missions. Assuming that these characteristics are Earth-like, settlers will carry equipment to: (i) extract minerals necessary to build the colony; (ii) produce energy, which can be solar or nuclear; and (iii) produce fuels or other sources of energy for use by surface vehicles. They will bring enough seed stocks to produce agricultural commodities before the colony becomes autonomous. Lastly, they will bring 3D printers to manufacture equipment and other products necessary for the colony survival until the return of the spaceship from Earth.

Mission's financing may follow a similar pattern held in the colonization of Massachusetts[46]. Let's see how it works. Individuals (the settlers), tired of living on the overcrowded Earth, look to move to the newfound planet. The settlers then petition the United Nations, which will be more likely renominated United Planets, to govern the colony according to a statute governed by its laws. They search for a group of wealthy individuals to finance the expedition through a new company (Deep Space Company or DSC). Remember that the space economy has extended to Mars and the asteroid belt, and Earth is

---

[45]A population size of 50 corresponds to an inbreeding rate of one percent per generation. This is half the maximum rate tolerated by domestic animal breeders. The population size of 500 balances the rate of gain in genetic variation with the rate of loss due to genetic drift.

[46]The Pilgrim Fathers were a small group of English Puritans. Most of them were farmers, poorly educated and without social or political standing. In 1617, discouraged by economic difficulties and their inability to secure civil autonomy, the congregation voted to emigrate to America. Unable to finance the costs of the voyage, they organized a joint-stock venture. They negotiated a financial agreement with a group of London businessmen, who provided the capital. The ship began its historic voyage in 1620 with 102 passengers. See Labaree B. (1979).

immensely richer than it is today. In exchange the settlers will grant DSC the rights of mineral exploitation on the newfound planet.

DSC finances the expedition for an amount equal to 55 % of the fuel and spaceship costs, for example 3.6 trillion dollars. The settlers, through crowd financing, buy the equipment and more. DSC deposits the two trillion dollars with an international public bank, such as the World Bank, for the mission's duration. Interest rate is four percent per year. The mission last thirty years in the spaceship's reference frame. Taking into account time dilation, time passed on Earth is seventy years. Thus the end value of the deposit will be around thirty one trillion dollars. DSC emits zero-coupon bonds. A zero-coupon bond is bought at a price lower than its nominal value, which is paid in full at expiration. The maturity of these zero-coupon bonds is seventy years. DSC decides that the price of zero-coupon bonds is 1.6 trillion dollars. Since these bonds are free of default risk due to the guaranteed bank deposit, the interest is also four per cent. The nominal value of the zero-coupon bonds is around twenty eight trillion dollars.

The settlers are now ready to go, with young investors traveling with it. On arrival, the spaceship will remain in orbit until a beachhead is established on the newfound planet. Upon return of the spaceship to Earth, DSC pays off zero-coupon bonds, realizes three trillion dollars surplus, and own the spaceship. Since the time elapsed for young investors is thirty years, their return on investment is around 85%, excluding the value of any cargo they may bring back. On the basis of information and requests by the settlers, the spaceship would then return to the extrasolar planet bringing cargo and new settlers, starting trade between the two planets.

Make a leap in time. Several individuals in the new colony leave the planet for another one distant five/fifteen light years. Over the centuries, humankind will have populated part of the galaxy. Man will inhabit semi-isolated islands in space, which will increase diversity, both cultural and genetic. Astronauts will form a separate civilization and connect these islands by guaranteeing trade between them. Time dilation helps them to lose emotional and other personal ties with inhabitants of various other islands in space. The special theory of relativity imposes these limits.

7.4 Overcoming the limits imposed by special relativity: myth or reality?

Limits imposed by the special theory of relativity are: (i) nothing travels at a speed greater than light; and (ii) clocks on board spaceships traveling at significant fractions of light's speed slow down compared to the cocks of observers at rest. Is it possible to overcome these limits? An answer is given by the general theory of relativity. This theory allows space-time warping. This is no fantasy but results from science and verified by astronomical observations[47]. The general theory of relativity requires being more precise in assertions about motion. Instead of saying that nothing can exceed the speed of light, it says that spaceships traveling in space cannot exceed it. If space-time is warped, this limit no longer applies. If space-time is locally deformed, it shrinks before and expands behind spaceships. Spaceships are pushed forward with space-time as surfboards ride on waves. They never travel locally at speeds greater than light since light is transported with the space-time expansion waves.

Miguel Alcubierre[48] worked on a theoretical demonstration. He proposed one space-time configuration in which a spaceship travels between two points in arbitrarily short time. Other studies[49] showed the theoretical possibility of distorting space-time geometry so that immense gravitational fields are not large near the spaceship and star bases. In this bubble, immense tidal forces due to gravity create no complications. Space can be nearly flat and clocks on the spaceship and star bases stay synchronized. But there is a huge gap between saying: "this violates none of the known laws of physics"; and "we have detected a bubble of space-time curvature in the real world." NASA is financing experiments to measure bubbles theorized by Alcubierre. Harold White did the first experiment to test Alcubierre-drive effects. This experiment used an interferometer to measure effects at nanometer scale[50]. So far the data are inconclusive. While experiments indicated an effect different from zero, external sources might have caused this difference. Other experiments are needed.

---

[47]The first test took place in 1919, when two British astronomical expeditions confirmed that the Sun's mass deforms space following the general theory of relativity. Other scientists confirmed that time is under gravity's influence. Clocks at high altitude slow down compared to clocks at sea level.

[48]Alcubierre A. (1994).

[49]See White H. G. et al. (2006).

[50]The original device proposed by White is an interferometer modified by using a laser beam. The beam is divided into two paths. The curvature of space causes a phase shift on each laser beams that should be detectable. Researchers tried to learn if space deformation caused by the energy of an electric field with high voltage could be detected. Plans are to increase sensitivity up to a hundredth of a wavelength and to create an oscillating field for more accurate results.

The good news ends here. The works mentioned above are related to the left members of Einstein equations[51], which, in a heuristic way, can be represented as follows:

(Space-time warp) = (mass-energy distribution)

Equations' left members set space-time geometry while equations' right members set mass-energy distribution. What mass-energy distribution is necessary to get space-time distortion needed for interstellar travel? To get this space-time configuration, Alcubierre's proposed mechanism implies a negative energy density. This is not surprising. To travel great distances in arbitrarily short time, solutions of general relativity equations imply that matter must gravitationally repel other matter. A theorem of the relativity's theory states that this condition is equivalent to the energy of matter being negative.

At present, we do not understand why negative energy exists or what was its origin. At macro level, observations of the rapid expansion of the universe[52] confirmed negative energy's existence. This finding is counter-intuitive since gravity exerts an attraction and should decelerate any expansion. The acceleration can be explained on the assumption that 70 per cent of the universes total energy is negative and 30% is matter. At microscopic level, quantum theory informs us that space is not empty. Quantum mechanics and the special theory of relativity imply that space's distribution of energy is negative. Merging quantum mechanic's vacuum condensation with general relativity results in negative energy wiping out the universe a few moments after the Big Bang. The universe would then be greatly different. This puzzle remains one of the crucial problems of theoretical physics. It suggests our incomplete understanding of the vacuum and/or of gravity.

Many theoretical physicists are working on a quantum theory of gravity. If such a theory is found, it may answer the questions on why negative energy exists and what is its origin. It would offer ways to produce negative energy. If this last statement seems crazy, think about electricity. Before Coulomb, Faraday and Maxwell, scientists considered electricity and magnetism two separate phenomena. No one had the faintest idea how to produce electricity. After Maxwell's four equations merged electricity and magnetism together, we learned to produce electricity on an industrial scale. The second industrial

---

[51]The Einstein formula is the following: $R_{\mu\nu}$-1/2 $Rg_{\mu\nu}$ + $\Lambda g_{\mu\nu}$ = $8\pi G/c^4$ $T_{\mu\nu}$, where $R\mu\nu$ is the Ricci curvature tensor, R is the scalar curvature, $g\mu\nu$ is the metric tensor, Λ is the cosmological constant, G is Newton's gravitational constant, c is the speed of light in vacuum, and $T\mu\nu$ is the stress–energy tensor.

[52]Astronomers used the stellar explosions or supernovas in distant galaxies to calculate their distances. When they measured the speed of recession of those galaxies, they discovered, to their surprise, that the universe's expansion is accelerating.

revolution took place in the second half of the 19th century. The inventions of communication devices (i.e., telegraph, radio, and telephone) revolutionized our ways of living and doing business. These developments bear witness of the importance of understanding key physical phenomena systematically and theoretically for the development of technologies and their effects on human civilization.

Even if someone elaborates a quantum theory of gravity, the energy needed to adapt space-time configurations for interstellar travels is extremely large. The equations of the general theory of relativity incorporate a constant that stipulates the elasticity of space-time. Einstein determined its value[53]. He found this coefficient equal to two times ten to the minus forty-eight (2 x 10-48). The inelasticity of space-time, given by the inverse of this ratio, is extremely large. Hence, the energy required to distort space-time is large. The question is whether it is possible to achieve the energy needed for interstellar flight. Many physicists believe interstellar travel, using space-time distortion, is practically impossible. But one thing is to be skeptical, another is to keep one's minds open. Eminent theorists[54] have argued on many occasions against phenomena proposed by general relativity. Later on, they reneged on their statements due to hard contrary evidence.

---

[53]Einstein assumed that, with approximation, these equations restore Newtonian theory starting from the gravitational potential. He found that the coefficient of elasticity of space-time is equal to $8\pi G/c^4$, where G is Newton's constant and c is the speed of light.

[54]In the thirties J.R. Oppenheimer presented a paper on black holes at a conference in Brussels. He hypothesized that objects known as neutron stars could not exceed two times the solar mass without collapsing into a black hole. An equally eminent theoretical physicist, J.A. Wheeler, argued this that this result is impossible, since the laws of physics protect the heavenly bodies from such an absurd end. Ten years later, Wheeler had to renege on this statement, and, by curious irony, it was he who gave the name blacks holes to these objects.

Bibliography

Alcubierre A.: *"The warp drive: hyper-fast travel within general relativity".* Classical and Quantum Gravity 11 (5), pp. 73–77, 1994.

Anglada G. et al.: *"Alma Discovery of Dust Belts around Proxima Centauri"*, The Astrophysical Journal Letters, forthcoming 2017.

Anglada G. et al.: *"Two planets around Kapteyn's star: a cold and a temperate super-Earth orbiting the nearest halo red dwarf"*, Monthly Notices of the Royal Astronomical Society Letters, Volume 443, Issue 1, pp. 89-93, 2014.

Bailey J. et al.: *"A Jupiter-like Planet Orbiting the Nearby M Dwarf GJ832"*, The Astrophysical Journal. 690: 743–747, 2008.

Bussard R. W.: *"Galactic Matter and Interstellar Flight"*, Astro Acta, 6, pp. 179-194, 1960.

Cowie L.: *"Spotlight on The Ages Of Exploration and Discovery"*, ed. Wayland Publishers, Sussex, England, 1985.

Greaves J. S. et al.: *"The debris disc around tau Ceti: a massive analogue to the Kuiper Belt"*. Monthly Notices of the Royal Astronomical Society. 351 (3): L54–L58, 2004.

Backman D. et al.: *"Epsilon Eridani's Planetary Debris Disk: Structure and Dynamics Based on Spitzer and Caltech Submillimeter Observatory Observations".* The Astrophysical Journal. 690 (2): 1522–1538, 2009.

De Cecco M.: *"Money and Empire: The International Gold Standard 1890-1914"*, Basil Blackwell, London 1973.

Dumusque X. et al.: *"An Earth-mass planet orbiting α Centauri B"*, Nature, 491, 207–211, 2012.

Forward R.: *"Antiproton Annihilation Propulsion"*, Journal of Propulsion and Power, Vol.1, No. 5, pp. 370-374, 1985.

Forward R.: *"Indistinguishable from Magic"*, Amazon Books, 1995.

Frisbee R.: *"How to build an antimatter rocket for interstellar missions"*, AIAA Paper 4676, 39th Conference, Huntsville, Alabama, July 2003.

Guedes J. M. et al.: *"Formation and Detectability of Terrestrial Planets around α Centauri B"*, The Astrophysical Journal, Volume 679, Issue 2, article id. 1582-1587, 2008

Kenneth A.: *"Trade, Plunder and Settlement: Maritime Enterprise and the Genesis of the British Empire, 1480–1630"*, ed. Cambridge University Press, 1984.

Kervella, et al.: *"Proxima's orbit around Alpha Centauri"*, Astronomy & Astrophysics 598, L7 (2017)

Kirck P. V.: *"On the Road of the Winds: An Archaeological History of the Pacific Islands before European Contact"*, ed. University of California Press, Berkeley, 2000.

Ingstad H. et al.: *"The Viking: The Excavation of a Norse Settlement in L'Anse Aux Meadows, Newfoundland"*, ed. Checkmark Books, 2001.

Labaree B.: *"Colonial Massachusetts: a History"*, KTO Press, Millwood, NY, 1979.

Lewis R.A. et al.: *"Antiproton-Catalyzed Microfission/Fusion Space Propulsion Systems for Exploration of the Outer Solar System and Beyond"*, Space Technology and Applications International Forum, January 26-30, 1997, Albuquerque, New Mexico, CONF-970115, ed. M. El-Genk, American Institute of Physics, New York.

Santerne A. et al.: *"SOPHIE Velocimetry of Kepler Transit Candidates"*, Astronomy and Astrophysics, 575, p. 76, 2012.

Schmidt G. R. and al.: *"Antimatter Production for Near-term Propulsion Applications"* NASA Marshall Space Flight Center, 1998.

Torres A. M.: *"Tau Ceti May Have a Habitable Planet"*, Astrobiology Magazine. December 19, 2012.

M. Tuomi et al.: *"Signals embedded in the radial velocity noise: Periodic variations in the Tau Ceti velocities"*. Astronomy & Astrophysics, 2012.

Van de Kamp P.: *"Alternate dynamical analysis of Barnard's star"*, Astronomical Journal 74 (8): pp. 757, 1969.

White H. G. et al.: *"The Alcubierre Warp Space Drive in Higher Dimensional Space Time"*, in Proceedings of Space Technology and Applications Forum, University of New Mexico and American Institute of Physics, 2006.

Chapter 8

Final Remarks

"Two things fill the mind with ever-increasing wonder and awe, the more often and the more intensely the mind of thought is drawn to them: the starry heavens above me and the moral law within me."

Immanuel Kant

Skeptics on space exploration and colonization as a way to the long term survival of the human race are numerous. Many observers think that man does not belong to space and that whatever resources are dedicated to space programs should be directed to scientific research. In his essay[1] "*Escaping Earth: Human Spaceflight as Religion*", historian Roger Launius argues that enthusiasm for space can be viewed as a religion. In one place of his paper, he clearly points to the issue: "*Like those espousing the immortality of the human soul among the world's great religions... statements of humanity's salvation through spaceflight are fundamentally statements of faith predicated on no knowledge whatsoever.*" But they are losing sight that space exploration is not just for scientific research, but also for other goals - economic, commercial, cultural, and, in the very long term, the survival of the human race - and that man has a leading role in it.

Launius assertion that we have no knowledge whatsoever is too pessimistic. At the moment, we have the knowledge and the technologies to access space resources close to Earth, like the Moon and the near earth asteroids. Space mining could thus happen within the next ten to twenty years. Moving further beyond the Moon and the near earth asteroids is more uncertain. It requires technologies that are not yet mature and a better understanding of the physical toll on the body of long duration space fights, and how to protect and even

[1]See Launius R.: *"Escaping Earth: Human Spaceflight as Religion"*, Astropolitics, 11:45–64, 2013.

fix them[2]. It also call for large public sector investments for human exploration and for international cooperation. The matter-antimatter engine, allowing interstellar travel, is within present scientific and technological capabilities, but will call for technologies that we do not yet have. Interstellar traveling depends also on the discovery of a planet hospitable to life within fifteen light years from Earth. If that happens, and it is a big if, the incentives to invest in the production of antimatter at reasonable costs would increase, opening the door to robotic and human missions.

In 1964, the Russian astrophysicist Nicolai Kardashev[3] suggested that the technical progress of a civilization is directly related to the energy its citizens manipulate. According to this theory, culture (in the broadest sense) is a product of energy and technology. Through technology, energy is used, and, since the social and philosophical systems are expressions of technology, culture rests on and is determined by the energy used. Kardashev projected three levels or types of civilization: Type I: Planetary Culture; it uses all available resources making skillful use of the entire power in the world (in our case 10^{15} watts); Type II: Stellar Culture; such civilization exploits the full power of its star (in our case, around 10^{26} watts); and Type III: Galactic Culture; this civilization harnesses the power of a galaxy. It harnesses 100 billion times the energy produced by a civilization of type II, more or less the same order-of-magnitude jump separating type I and II civilizations.

At present, humanity uses all available natural resources utilizing most, but not all, of the power on Earth. Our civilization is at ground zero but close to a type I civilization. As power generation is a leading cause of air pollution and the single largest source of global warming emissions, the energy choices made during this pivotal moment will have huge consequences for human health, climate, and the economy for decades to come. Right now the world is moving toward a natural gas-dominated electricity system, but an over-reliance on natural gas has significant risks and is not a long-term solution to the world energy needs. Other energy sources, like wind, solar power, geothermal energy, wave and tidal, and nuclear fusion are available on Earth, but with the exception of the onset of nuclear electricity, the energy mix has, for the most part, been fairly constant for at least half a century. Shifts in energy

---

[2]A recent study by NASA (see NASA (2017)) found that being in space encourages methylation, which is the biological process of switching genes on and off. Although the study is yet to be completed, the preliminary results shed light on understanding the microlevel changes of the body in space. The detailed findings of the study, scheduled to be published in 2018, provide a stepping stone toward long-duration space exploration such as journeys to Mars.

[3]See Kardashev N. (1964).

systems have historically been a slow process, particularly when coupled to long-term infrastructure[4]. This may explain the slow and marginal progress the world has made in the transition to modern renewable, and is a challenge that must be acknowledged if the world is to achieve a large-scale shift in the global energy systems. Shifting our energy systems away from fossil fuels towards renewable technologies will require significant financial investment. In 2004, the world invested 47 billion dollars. By 2015, this had increased to 286 billion dollars, an increase of more than 600 percent. Investment has grown across all regions, but at significantly different rates. If this trend persists, renewable sources could well reliably and affordably provide up to forty percent of the world electricity by 2030, and eighty percent by 2050, allowing to reach a Type I civilization.

But, as mineral resources on Earth are critical or limited, powerful nations are struggling to control these resources for their survival. Moving to a Type II civilization would be a better answer to the mounting problems the world is facing in the future. Exploiting the natural resources of the solar system and the full power of the Sun is within present scientific and technological capabilities. It would enlarge the natural resource base here on Earth and contribute to mitigating climate change, thereby conducting to global economic growth and a more peaceful geopolitical environment. It would contribute to the broadening of thought that acquires momentum by seeking opportunities for all who claim them. Opening a new frontier in space to the private realms of society through international cooperation for the benefit of the entire human community should thus be on the political agenda of space-faring nations as an important tool to strengthen world political stability, economic growth and welfare, and to deepen human potentialities and the evolution of its consciousness.

Moving to a Type II civilization is however contingent on the geopolitical environment here on Earth and international cooperation. It would require a civilization in which human beings see themselves as inhabitants of a single planet and in which global governance is conducted on a cooperative basis. This would allow a reduction of global military expenditures that could partly be channeled to the funding of space programs. It is a difficult path. Humanity is still struggling to find ways to cooperate, and is far from an advanced level of civilization in which international relations are solely based on cooperation and not conflict. The world today is confronting four crisis: (i) globalization and technological innovations; (ii) the resurgence of nationalism worldwide, (iii) global problems, such as pressures of overpopulation on nat-

---

[4]See V. Smil (2017)

ural resources, and (iv) a shift in political power that in many countries is mixed up with the economic game.

The world is in a transition period shaped by globalization and rapid technological innovation. Globalization is primarily an economic process of integration. It is characterized by increasing economic interdependence of national economies across the world through a cross-border movement of goods, services, technology, and capital. Current globalization trends can be largely accounted for by developed economies integrating with less developed economies by means of foreign direct investment, the reduction of trade barriers as well as other economic reforms, and, in many cases, immigration. Depending on the paradigm, economic globalization can be viewed as either a positive or a negative phenomenon. On the positive side, as more money is poured into developing countries, there is a greater chance for the people in those countries to economically succeed and increase their standard of living[5]. Global competition encourages creativity and innovation and keeps prices for commodities and services in check.

On the negative side, outsourcing, while it provides jobs to a population in one country, takes away those jobs from another country, leaving many without opportunities. This is particularly true as favorable conditions extended initially to developing countries in international treaties were not reviewed once these countries achieved the status of developed industrial countries. At the same time, rapid technological innovations in advanced countries are creating new jobs but also destroying old ones. Many governments failed to address those problems. Some failed to understand that compensation and rehabilitation ought to be viewed as routine aspects of social policy in a world of rapid economic change. Others failed to understand that the income distribution problem ought to be looked at more in terms of income security and of easing transitions rather than of subsidizing outmoded patterns of activity. This is at the root of the malaise pervading part of the population in many advanced industrial countries. This malaise led to the resurgence of nationalism. Many have heard the siren song of nationalism saying that the nation state is the only organization capable to protect its members at the exclusion of the outsiders. Resurgence of nationalism is more dangerous today due to the proliferation of nuclear weapons.

The growing concerns of population growth are not limited to environmental degradation such as losses of biodiversity, soil depletion, and toxic rivers and oceans. The pressure from population will require more energy to secure their needs, as each individual added to population will consume food, water,

---

[5]A comprehensive data set on world increasing standards of living can be found in Max Roser website www.ourworldindata.org.

and energy. The pollutants such as greenhouse gas emissions from industry, transportation, agricultural activities, deforestation, and desertification are all associated with massive population growth. This will result, at least in the short-medium term, in increases in global warming, and in inequity between people through the scarcity of resources. It goes further through the risks it imposes of epidemics, resource war, terrorism, and deaths from violent climatic events. In addition, climate change will likely aggravate regional agricultural problems.Areas where the soil is fragile, because of intensive farming, water logging, and wind erosion, will typically be much more sensitive to climate change than soils that have been protected and allowed to replenish. The soils in substantial parts of many developing countries have marginal physical characteristics. The region most at risk from problems of food security is Sub-Saharan Africa, which is more sensitive to reduced rainfall, rainfall variability, and evaporation than other regions. Only around two percent of its cropland is irrigated, fifty percent is in arid or semi-arid areas, and much of its soil is very fragile. Limited or critical mineral resources can lead to resource wars and instability in the geopolitical environment.

Recent years have seen the rise of political powers mixed up in the economic game. Awesome technological and financial powers are concentrated in few hands in our societies. Few people representing the public interest seem capable of grasping these issues. As a consequence, many people have lost the ability to set their own agenda and/or to question those in authority. An "alternative-fact" politics has emerged in connection with political and/or economic agendas of different vested interests in our societies. It has normalized the denial of scientific evidence, from climate change to discoveries of biological (rejection of what biological sciences have to say about human origins, sexuality, and reproduction) and health (scientifically discredited fears that vaccines cause autism have led to anti-vaccination movements, endangering public health) sciences. Many have fallen back, without noticing, into superstition and darkness.

The coming together of the four crises is an explosive mixture that may lead to conflicts, and slow down the process of technological innovation. History suggests another consideration. It teaches that civilization moves in a cycle. From the experiences of Babylon, Summer, Egypt, Rome, and China, we know that civilizations eventually faltered. And, as history also reveals, several circumstances were behind those falterings. Natural disasters, corruption, social injustice, and constant wars were among the causes. The classical world crises provide a good example of how that society did not address the issue of justice, particularly that of slavery. Brilliant minds never challenged the political, economic and religious foundations of their society. They apparently questioned everything in the fields of science, but never questioned slavery. The abundance of cheap slave labor was likely behind the classical world's inability to recognize technology's potential to free people. Coupled with constant wars, this inability led to the demise of classical civilizations, which accomplished spectacular successes in science and understanding of the natural world.

Presently we are experiencing an upward technological cycle. But the coming together of the four crisis may cause the high level of technology to decline. We may survive disaster for a while, but ignorance and power may stop the upward technological cycle. Human folly and other vices and fears can change the course and results of the journey we have tried to imagine, postponing human space exploration and colonization of the solar system to an indeterminate future. Despite this note of pessimism, the situation may not be hopeless. Michael Tomasello[6] argues that Homo Heidelbergensis developed abilities to cooperate through mimicry and gesturing in the search for food. Later on, Homo Sapiens expanded cooperative capabilities through common cultural backgrounds centered on conventions and norms. Through these arguments, Tomasello says that humans can continue to develop because they can absorb and share knowledge.

If the Tomasello's arguments are correct, cross cultural studies, a realistic biology of the mind, advances in physics, information technology, genetics, neurobiology, engineering, and chemistry of materials are all challenging the basic assumptions of who and what we are, and of what it means to be human, thereby contributing to human evolving. But these forces for change are so complex and interconnected that they would call for an enhanced role of education to understand: why the world is changing and the role of science in shaping these changes; how other people and cultures feel about those changes; and what humanity has in common and what divides cultures, classes, and nations. It would call for a thorough debate within our societies whether values, norms, and organizations inherited from the past are adequate to face the problems humanity faces today and in the near future.

Although this book is not a primer of solutions to global developments leading to a type II civilization, it is important to emphasize one key element to prepare the global society to face the challenges of the twenty first century, i.e. the role of the nation state. The end of the Cold War saw the re-emergence of the nation state as the primary focus of international politics. The role of international organizations is superseded whenever the pursuit of national interests prevails. But one cannot maintain international order or peace if unfettered sovereignty of nation states is kept. The great challenge of the twenty first century is thus to overcome the paradigm of the competition for

---

[6]See Tomasello M.(2014).

power among nation-states. Does this call for the establishment of a super state with the power to direct every single nation? The answer is no. What is needed is a minimum solution, meaning that there should be a political power, not mixed up in the economic game, to set up rules to restrain single nations to harm their neighbors, and having the authority to enforce those rules. The advantage of this solution is that it does not invoke an instant global reform. It would only call for all nation states to limit their sovereignty in their own interest, as peace and prosperity are common values.

The form of the political power is debated. Some argue that federalism is this form of power, being an application of democracy to international affairs. Other argue that rebuilding civilization on a large scale is a difficult process. The reason that a federation is a hierarchical structure that achieves consensus in dialectics with other organizations present in the social body. It implies that the existence of world citizenship is having political and legal status. At present, it is hard to imagine the existence of such world citizenship. Instead, they claim that what is needed is the establishment of internationally binding agreements on subjects of interest to all nations, and that an over ambitious attempt to form a world government is not likely to succeed. It is a debate about "foedus" versus a series of contracts that are binding as recognized useful to all nations involved. Critics to the second proposal ask the following question: how a stable division of the world into independent states or group of states with armed forces strictly confined to the mission of defending their own territory can be achieved; and how and who will monitor and enforce the compliance of individual nations?

The debate is going on. Unequivocal answers do not yet exist. But one fact emerges from this debate. It is clear that federalism would be the form of political power in a type II civilization. But to reach this ultimate goal a series of intermediate steps involving international agreements are necessary. Hereafter attention will be given to an international agreement among space faring nations to cooperate in developing space exploration programs and in establishing an international legal and institutional framework for a peaceful development by the private sector of the space economy beyond the Earth orbit. At present, different national space agencies have the mandate to develop space programs. As human exploration of the solar system will require large financial resources that are unlikely sustainable by any single nation, hence collaboration between space agencies is needed. One question arises here. What is the most efficient way of establishing cooperation between these agencies without hindering competition? One possible answer to this question is the creation of an international institution open to all nations to promote scientific and technical progress, industrial competitiveness, and joint initiatives needed for the exploration and exploitation of space.

The development of off-world sources of critical minerals by the private sector would prevent the competition among powerful state in space and its militarization, and would create an opportunity to transform space activity from consumer of resources into sources of value. One question arises here. How to maximize the role of the private realms of society in pushing the space economy beyond earth orbit? A possible answer to this question comes from the establishment of an international legal and institutional framework, as competition among private companies in developing space mineral resources without recourse to force will require it. An important question refers to who will have the authority to enforce the laws. Will it be the single nations participating to the international agreement or an international institution? The first solution is a consent-based governance, while the second one is not consent-based, and delegates national jurisdiction to supranational institutions. A Type II civilization requires the second solution, as, under the first solution, a state member is not obliged to abide by the international law, unless it has expressly consented to a particular course of conduct.

Moving to a Type III civilization and interstellar exploration is too remote a possibility to make realistic risk analysis sensible. But one can safely say that a necessary condition is for humanity to be able to move to a Type II civilization. The discovery of a habitable planet/s within 5 to 15 light years from Earth is also a crucial driver for interstellar exploration and colonization. Continuation of investments in both terrestrial and space-based telescopes is necessary allowing eventually to discover such planets. Moving beyond the limits imposed by the special theory of relativity will require a science and a technology that do not yet exist. But as long as there are men and women who, for their personal joy, will continue to push forward the frontier of knowledge and others who courageously push the frontier of human presence into the unknown, what is considered today impossible could become possible.

In summary, through science and education, humanity could find a common cultural background around which to develop conventions and norms leading to a global society that is less confrontational and based on mutual respect and non-violence. Exploration of the solar system and building of space infrastructures would help in a peaceful exploitation of the solar system resources, thereby enlarging the resource base on Earth and eliminating one of the causes of friction among nations on Earth. Through science, humanity can identify habitable planets in the Milky Way, develop the technologies for interstellar traveling within the limits imposed by the special theory of relativity, opening the way to move to a Type III civilization, and eventually find ways to overcome those limits.

The exploration of space is thus not a religion, but one of the tools that can help facing some of the challenges confronting humanity both today and in the future, including the challenge to develop a value system that allows humanity to live peacefully together in a global society. This is not a utopia, but a search for a value system rather than a set of shared beliefs that would result in a world more varied, more interesting and more hopeful, and in the process to reach the stars. The road will be bumpy, as human behavior is influenced not only by rational motivations, but also by cultural issues, religious taboos, greed and other less pleasant impulses, in essence our value system. But, like Sisyphus, humanity will be able to rise again after inevitable failures, perfectly representing the thought of Camus[7], according to whom "*we must imagine Sisyphus happy*".

I want to close this book with the words of Johann Scheffler. He was a German physician, also known as a mystic poet. Born in Silesia during the Thirty Years' War, he received a Lutheran education before studying medicine in Strasbourg, Leiden, and Padua. After his return to Silesia, he converted to Catholicism and adopted the name of Angelus Silesious, the name that appears in his book "The Cherubic Wanderer"[8]. This is a collection of two thousand epigrams that combine Baroque exuberance with the introspection typical of Nordic mysticism. Many of these epigrams have invited writers and poets to reflection, but there is one that is particularly appropriate to the present context:

> "Freund es ist auch genug. Jm fall du mehr wilt lesen/
>
> So geh und werde selbst die Schrifft und selbst das Wesen."

(Friend, it is now enough. Wouldst thou read more, go hence, and Become thyself the Writing and thyself the Sense.)

[7]A. Camus (1942)
[8]A. Silesious (1986)

Bibliography

Camus A.:"*Le Mythe de Sisyphe*", ed. Gallimard, Paris, 1942. English translation by Justin O'Brien, ed. Hamish Hamilton Limited, London, 1955.

Kardashev N.: "*The Transmission of Information by Extraterrestrial Civilizations*", Soviet Astronomy, 1964.

Launius R.: "*Escaping Earth: Human Spaceflight as Religion*", *Astropolitics*", 11:45-64, 2013.

NASA: "*Twin Study*", see https://www.nasa.gov/feature/exploring-the-ground-truth-nasa-s-twin-study-investigates-metabolites, 2017.

Shrady M.: "*Angelus Silesius: The Cherubic Wanderer*", Paulist Press, 1986.

Smil V.: "*Energy Transitions: Global and National Perspectives*", ed. Praeger, 2016.

Tomasello M.: "*A Natural History of Human Thinking*", Harvard University Press, 2014.

Index

A

Africa pp. 101
  Sub-Saharan Africa pp. 102
Alcubierre M. pp. 215
Asteroids pp. 68, 84
  Asteroids' mineral resources pp. 130, 133
  Asteroids' mining companies pp. 84
  Asteroids' mining equipment pp. pp. 136 to 138
  Economics of asteroid mining pp. 141, 146
  Near earth asteroids pp. 93, 121
Astronautics pp. 1, 4, 7, 16, 22, 64
  Hohman Transfer pp. 148
  Mars Trajectories pp. 174
  Rocket equation pp. 9, 24, 25, 139, 149, 176
  Relativistic rocket equation pp. 26
Astronomy pp. 194
  Habitable zone pp. 194
  Radial velocity pp. 194, 195
  Transit method pp. 195
  Transition time variation pp. 195
  Gravitational microlensing pp. 196
  Earth type planets pp. 198 to 205
  New Earth and space based telescopes pp. 206, 207

B

Bardi U. pp. 94
Bignami G. pp. 72
Blinkhorn M. pp. 21
Borowski S. pp. 170

C

China pp. 55
  Platinum Group Metals pp. 97
  Rare earth elements pp. 97
  Lauching industry pp. 59, 61
  Moon exploration pp. 127
  Space program pp. 65, 108
  Space budgets pp. 66
Clarke A. C. pp. 40
Commercial space industry pp. 53
  Commercial space services pp. 54
  Commercial space infrastructure pp. 56
  Costs of accessing space pp. 61
  Efficiency/costs gains of satellite applications pp. 78
  Launching industry pp. 59
  The dawn of commercial satellite activities pp. 43
Cosmism pp. 1, 3, 10
Craig H. W. pp. 72

E

Europe pp. 106
  European Space Agency pp. 63, 64

European Space Agency asteroid exploration missions pp. 134
European Space Agency budget pp 66
European Space Agency Moon Missions pp. 127
European Space Agency structure and policies pp. 107
Extraplanets within 15 light years from Earth pp. 194
around Alpha Centaury B pp. 198
around Bernard star pp. 200
around Epsilon Eridani pp. 201
around Gliese 832 pp. 205
around Kapteyn star pp. 203
around Proxima Centaury pp. 200
around Ross 128 pp. 202
around Tau Ceti pp. 202
around Wolf 1061 pp. 204

F

Forward R. pp. 192,193
Fyodorov N. pp.1

G

General relativity pp. 196
to travel great distances in arbitrarily short time pp. 216
Germany pp. 20
The Nazi rocket program pp. 21
Global issues pp. 89
Critical minerals supply pp. 96
Critical minerals demand pp. 97
Environmental problems pp. 99, 100
The limits to economically feasible mineral extraction pp. 94, 95
George L. E. pp. 175
Glushko V. P. pp.11, 12, 13
Goddard R. H. pp. 4
Goddard first groundbreaking work pp. 5
Goddard rocket experiments 6
Obstacles to change pp. 16, 17, 19, 22

I

India pp. 112
India space budget pp. 66
Launching industry .pp 59
The commercial space sector pp. 55
International cooperation pp. 103
Committee on Space Research pp. 75
International legal framework for space mining pp. 115
International space station pp. 76, 104
ITU agreement pp. 46
Outer Space Treaty pp. 47

J

Japan pp. 109
Asteroid exploration pp. 131
Japan space budget pp. 66
Moon exploration pp. 127
Space agency pp. 64
The commercial space sector pp. 58

K

Kardashev N. pp. 222
Kesler S. pp. 94
Kolodny N. G. pp. 3
Korolev S. P. pp. 10
  The early years pp. 11
  The gulag period pp. 12
  The 1950s 13, 30, 33, 34

L

Landau D. F. pp. 175

M

Mankins J. C. pp. 91, 92
Mars pp. 153
  Economic sustainability of a Mars permanent base pp. 165
  Mars colonization pp. 163
  Mars permanent base pp. 157
  Mars physical characteristics pp. 154
  Space elevator pp. 152
  Von Braun's Mars project pp. 48
Moon
  Apollo program pp. 35
  Apollo program's costs and benefits pp. 38
  Economic effects of the Apollo program pp. 38
  Lunar excursion module pp. 36
  Moon development companies pp. 84
  Moon mineral resources pp. 127
  Soviet Union Moon project pp. 38
Moore W. pp. 2

N

National Aeronautics and Space Administration pp. 35
  National Aeronautics and Space Administration and the Apollo program pp. 35, 37
  National Aeronautics and Space Administration budgets pp. 39
  National Aeronautics and Space Administration policy changes pp. 60

O

Oberth H. J. pp. 7, 11, 13, 16, 20, 22

P

Pierce J. R. pp. 43
Propulsion
  Chemical pp. 70, 170
  Electrical propulsion pp. 70, 91, 144
  Matter-antimatter propulsion, pp 192
  Nuclear fission pp. 71, 144, 170, 171
  Nuclear fusion pp. 70, 72, 189

R

Russia pp. 110
  International collaboration pp. 74
  International space station pp. 76
  Space budget of the Russian Federation pp. 66, 67

S

Schmidt G. R. pp. 191
Sommariva A. pp. 20
Soviet Union pp. 19
Cold war period pp. 29
Failed joint Moon project with the United States pp. 35
Military programs pp. 30
Space programs: the Sputnik pp. 33
Space programs: satellite pp. 40
Space based solar power pp. 91
Levelized costs of power sources pp. 92

T

Tomasello M. pp. 226
Tsander F. pp. 11
Tsiolkovsky K. E. pp. 1
Philosophical inquiries on the cosmos pp. 3
Theory of space flight pp. 2

U

United States pp. 17, 104
Commercial Space Launch Competitiveness Act pp. 116
Competition in space pp. 29, 45
International cooperation pp. 35, 46, 76
Military programs pp. 30, 31, 42
Open Sky policy pp. 54
Operation Paperclip pp. 16

V

von Braun W. pp. 10
Early years pp. 13
The VfR period pp. 14
The Nazi period pp. 15, 16
The American period pp. 30, 36, 39

W

White H. G. pp. 215

www.ingramcontent.com/pod-product-compliance
Lightning Source LLC
LaVergne TN
LVHW010721210926
848485LV00007B/1409/J